ISBN 978-0-266-99664-4
PIBN 10920807

LOGARITHMIC AND TRIGONOMETRIC

TABLES

BY

EDWARD A. BOWSER, LL.D.

PROFESSOR OF MATHEMATICS IN RUTGERS COLLEGE

BOSTON, U.S.A.

D. C. HEATH & CO., PUBLISHERS

1908

COPYRIGHT, 1895,
BY EDWARD A. BOWSER.

G

EXPLANATION OF THE TABLES.

PRELIMINARY NOTIONS.

1. The numerical calculations which occur in Trigonometry are very much abbreviated by the aid of logarithms. The rules for their use are as follows:

The logarithm of a product is equal to the sum of the logarithms of its factors.

The logarithm of a quotient is equal to the logarithm of the dividend minus the logarithm of the divisor.

The logarithm of any power of a number is equal to the logarithm of the number multiplied by the exponent of the power.

The logarithm of any root of a number is equal to the logarithm of the number divided by the index of the root.

For the investigations of these rules the student is referred to the "Treatise on Trigonometry," p. 87, or to works on Algebra.

The Common Logarithms (Briggs's) are the only ones used in extensive numerical calculations, and the only ones given in the following tables.

The common logarithm of a number is the exponent of that power of 10 which is equal to the number.

Thus, the logarithm of 100 is 2, because $10^2 = 100$. This is usually written $\log 100 = 2$. 10 is the *base* of the common system.

2. *A system of common logarithms* means the logarithms of all positive numbers to the base 10.

From the above definition of common logarithms, it follows that

$$10^0 = 1, \quad \therefore \log 1 = 0; \quad 10^{-1} = .1, \quad \therefore \log 0.1 = -1,$$
$$10^1 = 10, \quad \therefore \log 10 = 1; \quad 10^{-2} = .01, \quad \therefore \log 0.01 = -2,$$
$$10^2 = 100, \quad \therefore \log 100 = 2; \quad 10^{-3} = .001, \quad \therefore \log 0.001 = -3,$$
$$10^3 = 1000, \quad \therefore \log 1000 = 3; \text{ etc.} \qquad \log 0.0001 = -4, \text{etc.}$$

Hence, the logarithm of any number between 1 and 10 is some number between 0 and 1; i.e., 0 + a fraction; between 10 and 100 is some number between 1 and 2; i.e., 1 + a fraction, etc., etc.

Thus, it appears that the logarithm of any number greater than 1 is *positive*, and the logarithm of any positive number less than 1 is *negative;* and in general the logarithm of a number consists of two parts, an integral part and a decimal part.

The integral part is called the *characteristic* of · the logarithm, and may be *either positive or negative.*

The decimal part is called the *mantissa* of the logarithm, and is *always kept positive, in order that the mantissæ of the logarithms of all numbers expressed by the same digits in the same order may always be the same.*

3. It is evident from the above table that the characteristic can always be obtained by the following rule:

The characteristic of the logarithm of a number greater than unity is positive, and one less than the number of digits preceding the decimal point.

The characteristic of the logarithm of a number less than unity is negative, and one more than the number of ciphers immediately after the decimal point.

Thus, the characteristics of the logarithms of 3406, 340.6, 34.06, 3.406, .3406, .0003406, are respectively, 3, 2, 1, 0, − 1, − 4; the mantissæ are the same, being .53224.

Hence, log .0003406 = $\overline{4}$.53224, the minus sign being written *over the characteristic* to indicate that it only is negative, the mantissa being always positive.

4. In practice it is more common to avoid the use of negative characteristics by increasing them by 10, and then by allowing for it in the interpretation of the results.

Note. — It is only in rare cases that more than seven places of the mantissa are required; in general, four or five are sufficient; and it is only for the most accurate computations that six or seven are used.

5. *A table of logarithms* is a table by which the logarithm of any given number, or the number corresponding to any given logarithm, may be found.

TABLE I. LOGARITHMS OF NUMBERS (Pages 1-19).

6. This table gives the mantissæ of the logarithms of the natural numbers from 1 to 10009, calculated to five decimal places.* The characteristics are determined by the rule in Art. 3. On p. 1, both the characteristic and the mantissa are given.

7. *To find the logarithm of a given number.*

(1) *For a number of one, two, or three figures only.*

If the number has one or two figures, find it on page 1 in the column headed N. Then in the same horizontal line as the mumber, and in the next column headed Log, will be found its logarithm.

Thus, $\log 7 = 0.84510$; $\log 68 = 1.83251$.

If the number has three figures, find on one of the pages 2–19, in the column headed N, the given number. Then in the same horizontal line as the number, and in the next vertical column, which is headed 0, will be found the mantissa of its logarithm: prefix the characteristic by the rule in Art. 3.

Thus, $\log 415 = 2.61805$; $\log 94.8 = 1.97681$.

Note 1. — A dash under a terminal 5 indicates that the true value is less than 5. Thus the logarithm of 415 to seven decimal places is 2.6180481. If only five decimal places are required, we neglect the 81 and increase 4 to 5. If six decimal places are required, the 1 is neglected; thus the above logarithm is written 2.618048.

(2) *For a number of four figures.*

Find on one of the pages 2–19, in the column headed N, the first three figures of the given number. Then in the same horizontal line as the first three figures, and in the vertical column which has the fourth figure of the given number at the top, will be found the last three figures of the mantissa of the required logarithm, to which the first two figures in the nearest mantissa above, in the column headed 0, are to be prefixed; supply the characteristic by the rule in Art. 3.

Note 2. — To save space, only the last three figures of the mantissæ are given in the columns headed 0, 1, 2, 3, 4, 5, 6, 7, 8, 9, and the first two at intervals, in the column under L. When the first two figures are not

---

* With five decimal places the numbers will be correct to the one hundred-thousandth part of a unit, which is near enough for most practical applications.

given in any line, they are to be taken from the first line above containing them, unless the last three are preceded by a star *, in which case they are to be taken from the line immediately below.

Thus, (p. 13) log 6615 = 3.82053, log 67.36 = 1.82840, log 6.764 = 0.83020.

(3) *For a number of more than four figures.*
To find log 2845.672.
We find from the table on p. 5, as in (2),

$$\log 2845 = 3.45408$$
and
$$\log 2846 = 3.45423$$
$$\overline{\text{diff. for } 1 = 0.00015}$$

Thus, for an increase of 1 in the number there is an increase of .00015 in the logarithm.

Hence, *assuming* that the increase of the logarithm is proportional to the increase of the number, then *an increase in the number of .672 will correspond to an increase in the logarithm of* .672 × .00015 = .00010, *to the nearest fifth decimal place.*

Hence
$$\log 2845 = 3.45408$$
$$\text{diff. for } .672 = .00010$$
$$\overline{\therefore \log 2845.672 = 3.45418}$$

NOTE 3. — We assumed in this method that the increase in a logarithm is *proportional* to the increase in the number. Although this is not *strictly* true, yet in most cases it is sufficiently exact for practical purposes.

8. From the above work we have the following rule for a number of more than four figures :

Find the tabular mantissa of the first four significant figures of the number; subtract this mantissa from the next greater tabular mantissa; multiply the difference thus found by the remaining figures of the number, as a decimal; add the product to the mantissa of the first four figures, and prefix the proper characteristic.

NOTE 4. — The difference between any mantissa in the table and the mantissa of the next higher number of four figures, is called the *tabular difference;* and the corresponding *proportional parts* are placed in the column headed *P.P.* By means of this column of proportional parts the above multiplication is facilitated.

It will be seen that this difference between the logarithms of two consecutive numbers is not always the same; for instance, those in the upper part of p. 5 differ by .00018, while those in the middle and the lower parts differ by .00016 and .00014. In the column with the heading 15 we see the difference 9 corresponding to the figure 6, which implies that when the difference between the logarithms of two consecutive numbers is .00015, the increase in the *logarithm* corresponding to an increase of .6 in the *number* is .00009.

Thus, the mantissæ of the logarithms of 2845 and 2846 differ by .00015; therefore 15 is the *tabular difference.* Then in the proportional table for 15, we find

$$\frac{1}{10} \quad \text{the proportional part for} \quad 6 = 9.00$$
$$\frac{1}{10} \quad \text{"} \quad \text{"} \quad \text{"} \quad 7 = 1.05$$
$$\frac{1}{100} \quad \text{"} \quad \text{"} \quad \text{"} \quad 2 = 0.03$$
$$\text{"} \quad \text{"} \quad 672 = 10.08,$$

or 10 to the nearest integer, which agrees with the value above.

9. *To find the number corresponding to a given logarithm.*

By reversing the above operations, the number corresponding to a given logarithm may be found, as will be seen by the following example:

Find the number whose logarithm is 3.47384.

We find that this mantissa does not occur exactly in the table. We therefore take out the next smaller mantissa, .47378 (on p. 5), whose corresponding number is 2977, and the next greater mantissa .47392, whose corresponding number is 2978.

The difference between these two mantissæ = .00014.
The difference between the smaller and given mantissæ = .00006.

Thus, for an increase of 1 in the number, there is an increase of .00014 in the mantissa; hence for an increase of .00006 in the mantissa there will be an increase of $\frac{6}{14}$ of 1 in the number = .43.

Hence, the number corresponding = 2977.43.

From the above work we have the following rule:

Find the tabular mantissa next less than the given mantissa, and the corresponding number of four figures; divide the difference of these mantissæ by the tabular difference, annex the quotient to the first four figures of the number, and point off the result according to the characteristic.

NOTE. — The labor of division may be saved by using the table of proportional parts for 14, as follows:

Given mantissa	= .47384
mantissa of 2977	= .47378
diff. of mantissæ	= 6
proportional part for 4 =	5.6
	0.4
proportional part for .3 =	.42
	− 0.02

∴ number = 2977.43 . . .

10. *Arithmetic Complement.* By the *arithmetic complement* of the logarithm of a number, or briefly, the *cologarithm* of the number, is meant the remainder found by subtracting the logarithm from 10. To subtract one logarithm, *b*, from another, *a*, is the same as to add the cologarithm, $10 - b$, and then subtract 10 from the result.

Thus, $a - b = a + (10 - b) - 10.$

When one logarithm is to be subtracted from the sum of several others, it is more convenient to *add* its *cologarithm* to the sum, and reject 10. The advantage of using the cologarithm is that it enables us to exhibit the work in a more compact form.

The cologarithm is easily taken from the table mentally by subtracting the last significant figure on the right from 10, and all the others from 9.

TABLE II. LOGARITHMS OF SINES, TANGENTS, ETC.

11. This table (pages 21–82) contains the logarithms of the sines, tangents, cotangents, and cosines of all angles from 0° to 90°.

If the angle is less than 45°, we look for the name of the function and the number of degrees in the angle at the *top* of the page, and the minutes in the *left-hand* column.

If the angle is between 45° and 90°, we look for the name of the function and the number of degrees at the *bottom* of the page, and the minutes in the *right-hand* column. In each case the horizontal rows at the top of the pages go with the degrees at the top, and the horizontal rows at the bottom go with the degrees at the bottom.

On pp. 21–33 the minutes and each ten seconds are given in columns at the left and right, and the odd seconds are given in a horizontal row at the top and bottom of each page. On

pp. 34–82 the minutes are given in columns at the left and right-
and on pp. 34–43 each ten seconds is given in a horizontal row
at the top and bottom of each page.

It is sufficient to have tables which give the functions of
angles only in the first quadrant, since the functions of all angles
of whatever size can be reduced to functions of angles less than
90° (Art. 35).

12. Since the sines and cosines of all angles, the tangents
of angles less than 45°, and the cotangents of angles greater
than 45°, are *less than unity*, the logarithms of these functions are
negative. To avoid the inconvenience of using negative characteristics, 10 is added to the logarithms of all these functions before
they are entered in the table. The logarithms so increased are
called the *tabular logarithms* of the sine, tangent, etc.

Thus, log sin 27° 48' = 9.66875;
　　　　log tan 27° 48' = 9.72201;
　　　　log cot 70° 5' = 9.55910;
　　　　log cos 27° 48' = 9.94674.

13. *To find the logarithmic sine, tangent, etc., of a given angle.*

(1) *When the angle contains only degrees and minutes.*

In this case the logarithm is given immediately in the table.
Thus we find (pp. 56 and 57), the following:

　log sin 18° 38' = 9.50449.　　　log sin 71° 13' = 9.97623.
　log cot 19° 23' = 0.45367.　　　log tan 70° 51' = 0.45935.

(2) *When the angle contains degrees, minutes, and seconds.*

In this case we take out the logarithmic function for the
degrees and minutes, as in (1); the correction for the seconds
has to be calculated in the same manner as for the logarithms
of numbers (Art. 7). For this purpose, on pp. 44–82, the differences of the logarithmic sines and cosines for 1' are given in the
columns headed d. (*difference*), and those of the logarithmic tangents and cotangents in the columns headed c. d. (*common difference*). In the case of tangent and cotangent, only one column of
differences is necessary for both functions.

Ex. 1. To find log sin 18° 25' 35''.

Page 56, log sin 18° 25' = 9.49958. Tabular difference = 38.
Hence, diff. for 35'' = $\frac{35}{60} \times 35$ = 　22
　∴ log sin 18° 25' 35'' = 9.49980.

From this work, we have the following rule:

Find from the table the logarithmic function for the degrees and minutes, and the corresponding tabular difference; divide this difference by 60, multiply the quotient by the number of seconds, and add this correction, if the function is a sine or tangent; or subtract it, if the function is a cosine or cotangent.

NOTE. — The labor of multiplication and division may be saved by means of the tables of proportional parts given in the right margin, the use of which is similar to those given in the table of logarithms of numbers. The proportional parts between 0° and 1°, and 89° and 90°, are given for 1, 2, 3, etc., tenths of a second; and between 1° and 89° they are given for 1, 2, 3, etc., seconds.

Ex. 2. To find log tan 64° 35' 18".

Page 63,   log tan 64° 35' = 0.32313.   Tab. diff. = 33.
     Under diff. 33, P.P. for 10" =   5.5
                    P.P. for 8" =   4.4
          ∴ log tan 64° 35' 18" = 0.32323.

Ex. 3. To find log cos 35° 44' 49".

Page 73,   log cos 35° 44' = 9.90942.   Tab. diff. = 9.
     Under diff. 9, P.P. for 40" =   − 6.0
                   P.P. for 9" =   − 1.4
          ∴ log cos 35° 44' 49" = 9.90935.

NOTE. — Since the cosine diminishes as the angle increases, we *subtract* the proportional parts.

14. *To find the angle corresponding to a given logarithmic sine or tangent.*

By reversing the above operations, the angle corresponding to a given logarithmic function may be found, as will be seen by the following examples:

Ex. 1. Find the angle whose log sin = 9.81510.

We find that this mantissa does not occur exactly in the column of logarithmic sines. We therefore take out the next smaller logarithmic sine, 9.81505 (on p. 78), whose corresponding angle is 40° 47', and the tabular difference 14. The difference between this logarithm and the given one is 5. The difference for 1" is 14 ÷ 60 or .23; hence for an increase of 5 in the mantissa there will be an increase of $\frac{5}{.23}$ seconds in the angle = 21".

Hence, the angle corresponding = 40° 47' 21".

From this work we have the following rule:

Find the tabular logarithmic function next less than the given function, and the corresponding degrees and minutes; divide the difference of these logarithms by the difference for 1", and annex the quotient to the degrees and minutes.

NOTE 1. — The method for finding the angle corresponding to a given logarithmic *cosine* or *cotangent* is the same, except that we find the next *greater* tabular logarithmic function, instead of the next less.

NOTE 2. — The labor of division may be saved by using the tables of proportional parts.

Ex. 2. Find the angle whose log tan = 9.87258.

	Given log tangent = 9.87258	
Page 74,	log tan 36° 42' = 9.87238.	Tab. diff. = 26.
Difference of logarithms	= 20	
Under tab. diff. 26, P.P. for 40" =	17.3	
	2.7	
Under tab. diff. 26, P.P. for 6" =	2.6	
	.1	

∴ required angle = 36° 42' 46".

Ex. 3. Find the angle whose log cos = 9.27235.

	Given log cosine = 9.27235	
Page 48,	log cos 79° 12' = 9.27273.	Tab. diff. = 67.
Difference of logarithms	= 38	
Under tab. diff. 67, P.P. for 30" =	33.5	
	4.5	
Under tab. diff. 67, P.P. for 4" =	4.5	

∴ required angle = 79° 12' 34".

15. SINES, ETC., OF SMALL ANGLES.

Ex. 1. To find log sin 0° 45' 37".28.

Page 30,   log sin 0° 45' 37" = 8.12284.   Tab. diff. = 16.

Hence, diff. for 0".28 = 16 × .28 =   4.48

∴ log sin 0° 45' 37".28 = 8.12288

NOTE. — The tables of proportional parts may be used, as explained in Art. 13.

Ex. 2. To find log tan 0° 54′ 27″.68.

Page 33,   log tan 0° 54′ 27″ = 8.19976.   Tab. diff. = 13.
    Under diff. 13, P.P. for 0″.6 =   7.8
                    P.P. for   .08 =   1.04
    ∴ log tan 0° 54′ 27″.68 = 8.19985.

Ex. 3. To find log cos 89° 22′ 35″.63.

Page 28,   log cos 89° 22′ 35″ = 8.03678.   Tab. diff. = 19.
                    P.P. for 0″.6 =   − 11.4
                    P.P. for   .03 =   − 0.57
    ∴ log cos 89° 22′ 35″.63 = 8.03666.

Ex. 4. To find log sin 4° 36′ 58″.6.

Page 40,   log sin 4° 36′ 50″ = 8.90548.   Tab. diff. = 26.
                    P.P. for 8″   =   20.8
                    P.P. for 0.6   =   1.56
    ∴ log sin 4° 36′ 58″.6 = 8.90570.

Ex. 5. To find log tan 5° 14′ 46″.4.

Page 43,   log tan 5° 14′ 40″ = 8.96279.   Tab. diff. = 23.
                    P.P. for 6″   =   13.8
                    P.P. for 0.4   =   0.92
    ∴ log tan 5° 14′ 46″.4 = 8.96294.
Page 42,   log cos 5° 38′ 32″.8 = 9.99789.

Ex. 6. To find log cot 85° 45′ 23″.7.

Page 41,   log cot 85° 45′ 20″ = 8.87049.   Tab. diff. = 29.
                    P.P. for 3″   =   − 8.7
                    P.P. for 0.7   =   − 2.03
    ∴ log cot 85° 45′ 23″.7 = 8.87038.

Note. — When the logarithmic function is given, the angle may be found by reversing the above operations, as in Art. 14.

TABLE III. THE NATURAL* TRIGONOMETRIC FUNCTIONS.

16. This table (pp. 84–87) contains the natural sines, tangents, etc., of angles from 0° to 90°, at intervals of 10′, calculated to four places of decimals. If greater accuracy is required it may be obtained by the proportional parts.

---

* A table which gives the values of the trigonometric sines, cosines, etc., is called a *table of natural trigonometric functions.*

17. *To find the sine, tangent, etc., of a given angle.*

If the *sine* or *tangent* is required, we look for the degrees in the *left-hand* column, and the minutes at the *top* of the page. If the *cosine* or *cotangent* is required, we look for the degrees in the *right-hand* column, and the minutes at the *bottom* of the page. The use is similar to that of the table of logarithmic functions, as may be seen by the following examples:

Ex. 1. Find the sine of 28° 14′.

Page 84,   sin 28° 10′ = 0.4720.   Tab. diff. = 26.
   Under diff. 26, P.P. for 4′ =   10.4
                ∴ sin 28° 14′ = 0.4730.

Ex. 2. Find the cotangent of 81° 49′.
Page 86,   cot 81° 40′ = 0.146$\underline{5}$.   Tab. diff. = 30.
   Under diff. 30, P.P. for 9′ =   − 27
                ∴ cot 81° 49′ = 0.1438.

18. *To find the angle corresponding to a given sine, tangent, cosine, or cotangent.*

Ex. 1. Find the angle whose cosine is .4585.
                Given cosine = 0.4585
Page 84,   cos 62° 40′ = 0.4592.   Tab. diff. = 26.
                diff. =   7
Under tab. diff. 26, P.P. for 2′ =   5.2
                                          1.8
      "   "   "   " .7′ =   1.8
∴ required angle = 62° 42′.7.

TABLE I.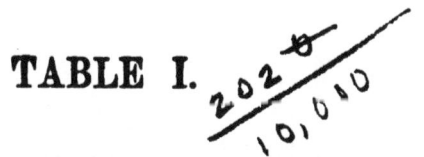

COMMON LOGARITHMS OF NUMBERS.

From 1 to 10009.

N.	Log.	N.	Log.	N.	Log.	N.	Log.	N.	Log.	N.	Log.
0	—∞	20	1.30 103	40	1.60 206	60	1.77 815	80	1.90 309		
1	0.00 000	21	1.32 222	41	1.61 278	61	1.78 533	81	1 90 849		
2	0.30 103	22	1.34 242	42	1.62 325	62	1.79 239	82	1.91 381		
3	0.47 712	23	1.36 173	43	1.63 347	63	1.79 934	83	1.91 908		
4	0.60 206	24	1.38 021	44	1.64 345	64	1.80 618	84	1.92 428		
5	0.69 897	25	1.39 794	45	1.65 321	65	1.81 291	85	1.92 942		
6	0.77 815	26	1.41 497	46	1.66 276	66	1.81 954	86	1.93 450		
7	0.84 510	27	1.43 136	47	1.67 210	67	1.82 607	87	1.93 952		
8	0.90 309	28	1.44 716	48	1.68 124	68	1.83 251	88	1.94 448		
9	0.95 424	29	1.46 240	49	1.69 020	69	1.83 885	89	1.94 939		
10	1.00 000	30	1.47 712	50	1.69 897	70	1.84 510	90	1.95 424		
11	1.04 139	31	1.49 136	51	1.70 757	71	1.85 126	91	1.95 904		
12	1.07 918	32	1.50 515	52	1.71 600	72	1.85 733	92	1.96 379		
13	1.11 394	33	1.51 851	53	1.72 428	73	1.86 332	93	1.96 848		
14	1.14 613	34	1.53 148	54	1.73 239	74	1.86 923	94	1.97 313		
15	1.17 609	35	1.54 407	55	1.74 036	75	1.87 506	95	1.97 772		
16	1.20 412	36	1.55 630	56	1.74 819	76	1.88 081	96	1.98 227		
17	1.23 045	37	1.56 820	57	1.75 587	77	1.88 649	97	1.98 677		
18	1.25 527	38	1.57 978	58	1.76 343	78	1.89 209	98	1.99 123		
19	1.27 875	39	1.59 106	59	1.77 085	79	1.89 763	99	1.99 564		
20	1.30 103	40	1.60 206	60	1.77 815	80	1.90 309	100	2.00 000		

N.	L. 0	1	2	3	4	5	6	7	8	9
100	00 000	043	087	130	173	217	260	303	346	389
101		432	475	518	561	604	647	689	732	775	817
102		860	903	945	988	*030	*072	*115	*157	*199	*242
103	01 284	326	368	410	452	494	536	578	620	662
104		703	745	787	828	870	912	953	995	*036	*078
105	02 119	160	202	243	284	325	366	407	449	490
106		531	572	612	653	694	735	776	816	857	898
107		938	979	*019	*060	*100	*141	*181	*222	*262	*302
108	03 342	383	423	463	503	543	583	623	663	703
109		743	782	822	862	902	941	981	*021	*060	*100
110	04 139	179	218	258	297	336	376	415	454	493
111		532	571	610	650	689	727	766	805	844	883
112		922	961	999	*038	*077	*115	*154	*192	*231	*269
113	05 308	346	385	423	461	500	538	576	614	652
114		690	729	767	805	843	881	918	956	994	*032
115	06 070	108	145	183	221	258	296	333	371	408
116		446	483	521	558	595	633	670	707	744	781
117		819	856	893	930	967	*004	*041	*078	*115	*151
118	07 188	225	262	298	335	372	408	445	482	518
119		555	591	628	664	700	737	773	809	846	882
120		918	954	990	*027	*063	*099	*135	*171	*207	*243
121	08 279	314	350	386	422	458	493	529	565	600
122		636	672	707	743	778	814	849	884	920	955
123		991	*026	*061	*096	*132	*167	*202	*237	*272	*307
124	09 342	377	412	447	482	517	552	587	621	656
125		691	726	760	795	830	864	899	934	968	*003
126	10 037	072	106	140	175	209	243	278	312	346
127		380	415	449	483	517	551	585	619	653	687
128		721	755	789	823	857	890	924	958	992	*025
129	11 059	093	126	160	193	227	261	294	327	361
130		394	428	461	494	528	561	594	628	661	694
131		727	760	793	826	860	893	926	959	992	*024
132	12 057	090	123	156	189	222	254	287	320	352
133		385	418	450	483	516	548	581	613	646	678
134		710	743	775	808	840	872	905	937	969	*001
135	13 033	066	098	130	162	194	226	258	290	322
136		354	386	418	450	481	513	545	577	609	640
137		672	704	735	767	799	830	862	893	925	956
138		988	*019	*051	*082	*114	*145	*176	*208	*289	*270
139	14 301	333	364	395	426	457	489	520	551	582
140		613	644	675	706	737	768	799	829	860	891
141		922	953	983	*014	*045	*076	*106	*137	*168	*198
142	15 229	259	290	320	351	381	412	442	473	503
143		534	564	594	625	655	685	715	746	776	806
144		836	866	897	927	957	987	*017	*047	*077	*107
145	16 137	167	197	227	256	286	316	346	376	406
146		435	465	495	524	554	584	613	643	673	702
147		732	761	791	820	850	879	909	938	967	997
148	17 026	056	085	114	143	173	202	231	260	289
149		319	348	377	406	435	464	493	522	551	580
150		609	638	667	696	725	754	782	811	840	869

P. P.

	44	43	42
1	4,4	4,3	4,2
2	8,8	8,6	8,4
3	13,2	12,9	12,6
4	17,6	17,2	16,8
5	22,0	21,5	21,0
6	26,4	25,8	25,2
7	30,8	30,1	29,4
8	35,2	34,4	33,6
9	39,6	38,7	37,8

	41	40	39
1	4,1	4,0	3,9
2	8,2	8,0	7,8
3	12,3	12,0	11,7
4	16,4	16,0	15,6
5	20,5	20,0	19,5
6	24,6	24,0	23,4
7	28,7	28,0	27,3
8	32,8	32,0	31,2
9	36,9	36,0	35,1

	38	37	36
1	3,8	3,7	3,6
2	7,6	7,4	7,2
3	11,4	11,1	10,8
4	15,2	14,8	14,4
5	19,0	18,5	18,0
6	22,8	22,2	21,6
7	26,6	25,9	25,2
8	30,4	29,6	28,8
9	34,2	33,3	32,4

	35	34	33
1	3,5	3,4	3,3
2	7,0	6,8	6,6
3	10,5	10,2	9,9
4	14,0	13,6	13,2
5	17,5	17,0	16,5
6	21,0	20,4	19,8
7	24,5	23,8	23,1
8	28,0	27,2	26,4
9	31,5	30,6	29,7

	32	31	30
1	3,2	3,1	3,0
2	6,4	6,2	6,0
3	9,6	9,3	9,0
4	12,8	12,4	12,0
5	16,0	15,5	15,0
6	19,2	18,6	18,0
7	22,4	21,7	21,0
8	25,6	24,8	24,0
9	28,8	27,9	27,0

N.	L. 0	1	2	3	4	5	6	7	8	9	P. P.

N.	L. 0	1	2	3	4	5	6	7	8	9
150	17 609	638	667	696	725	754	782	811	840	869
151	898	926	955	͵984	͵013	*041	*070	*099	*127	*156
152	18 184	213	241	270	298	327	355	384	412	441
153	469	498	526	554	583	611	639	667	696	724
154	752	780	808	837	865	893	921	949	977	*005
155	19 033	061	089	117	145	173	201	229	257	285
156	312	340	368	396	424	451	479	507	535	562
157	590	618	645	673	700	728	756	783	811	838
158	866	893	921	948	976	*003	*030	*058	*085	*112
159	20 140	167	194	222	249	276	302	330	358	385
160	412	439	466	493	520	548	575	602	629	656
161	683	710	737	763	790	817	844	871	898	925
162	952	978	*005	*032	*059	*085	*112	*139	*165	*192
163	21 219	245	272	299	325	352	378	405	431	458
164	484	511	537	564	590	617	643	669	696	722
165	748	775	801	827	854	880	906	932	958	985
166	22 011	037	063	089	115	141	167	194	220	246
167	272	298	324	350	376	401	427	453	479	505
168	531	557	583	608	634	660	686	712	737	763
169	789	814	840	866	891	917	943	968	994	*019
170	23 045	070	096	121	147	172	198	223	249	274
171	300	325	350	376	401	426	452	477	502	528
172	553	578	603	629	654	679	704	729	754	779
173	805	830	855	880	905	930	955	980	*005	*080
174	24 055	080	105	130	155	180	204	229	254	279
175	304	329	353	378	403	428	452	477	502	527
176	551	576	601	625	650	674	699	724	748	773
177	797	822	846	871	895	920	944	969	993	*018
178	25 042	066	091	115	139	164	188	212	237	261
179	285	310	334	358	382	406	431	455	479	503
180	527	551	575	600	624	648	672	696	720	744
181	768	792	816	840	864	888	912	935	959	983
182	26 007	031	055	079	102	126	150	174	198	221
183	245	269	293	316	340	364	387	411	435	458
184	482	505	529	553	576	600	623	647	670	694
185	717	741	764	788	811	834	858	881	905	928
186	951	975	998	*021	*045	*068	*091	*114	*138	*161
187	27 184	207	231	254	277	300	323	346	370	393
188	416	439	462	485	508	531	554	577	600	623
189	646	669	692	715	738	761	784	807	830	852
190	875	898	921	944	967	989	*012	*035	*058	*081
191	28 103	126	149	171	194	217	240	262	285	307
192	330	353	375	398	421	443	466	488	511	533
193	556	578	601	623	646	668	691	713	735	758
194	780	803	825	847	870	892	914	937	959	981
195	29 003	026	048	070	092	115	137	159	181	203
196	226	248	270	292	314	336	358	380	403	425
197	447	469	491	513	535	557	579	601	623	645
198	667	688	710	732	754	776	798	820	842	863
199	885	907	929	951	973	994	*016	*038	*060	*081
200	30 103	125	146	168	190	211	233	255	276	298
N.	L. 0	1	2	3	4	5	6	7	8	9

P. P.

	29	28
1	2,9	2,8
2	5,8	5,6
3	8,7	8,4
4	11,6	11,2
5	14,5	14,0
6	17,4	16,8
7	20,3	19,6
8	23,2	22,4
9	26,1	25,2

	27	26
1	2,7	2,6
2	5,4	5,2
3	8,1	7,8
4	10,8	10,4
5	13,5	13,0
6	16,2	15,6
7	18,9	18,2
8	21,6	20,8
9	24,3	23,4

	25
1	2,5
2	5,0
3	7,5
4	10,0
5	12,5
6	15,0
7	17,5
8	20,0
9	22,5

	24	23
1	2,4	2,3
2	4,8	4,6
3	7,2	6,9
4	9,6	9,2
5	12,0	11,5
6	14,4	13,8
7	16,8	16,1
8	19,2	18,4
9	21,6	20,7

	22	21
1	2,2	2,1
2	4,4	4,2
3	6,6	6,3
4	8,8	8,4
5	11,0	10,5
6	13,2	12,6
7	15,4	14,7
8	17,6	16,8
9	19,8	18,9

N.	L. 0	1	2	3	4	5	6	7	8	9	P. P.
200	30 108	125	146	168	190	211	233	255	276	298	
201		320	341	363	384	406	428	449	471	492	514
202		535	557	578	600	621	643	664	685	707	728
203		750	771	792	814	835	856	878	899	920	942
204		963	984	*006	*027	*048	*069	*091	*112	*133	*154
205	31 175	197	218	239	260	281	302	323	345	366	
206		387	408	429	450	471	492	513	534	555	576
207		597	618	639	660	681	702	723	744	765	785
208		806	827	848	869	890	911	931	952	973	994
209	32 015	035	056	077	098	118	139	160	181	201	
210		222	243	263	284	305	325	346	366	387	408
211		428	449	469	490	510	531	552	572	593	613
212		634	654	675	695	715	736	756	777	797	818
213		838	858	879	899	919	940	960	980	*001	*021
214	33 041	062	082	102	122	143	163	183	203	224	
215		244	264	284	304	325	345	365	385	405	425
216		445	465	486	506	526	546	566	586	606	626
217		646	666	686	706	726	746	766	786	806	826
218		846	866	885	905	925	945	965	985	*005	*025
219	34 044	064	084	104	124	143	163	183	203	223	
220		242	262	282	301	321	341	361	380	400	420
221		439	459	479	498	518	537	557	577	596	616
222		635	655	674	694	713	733	753	772	792	811
223		830	850	869	889	908	928	947	967	986	*005
224	35 025	044	064	083	102	122	141	160	180	199	
225		218	238	257	276	295	315	334	353	372	392
226		411	430	449	468	488	507	526	545	564	583
227		603	622	641	660	679	698	717	736	755	774
228		793	813	832	851	870	889	908	927	946	965
229		984	*003	*021	*040	*059	*078	*097	*116	*135	*154
230	36 173	192	211	229	248	267	286	305	324	342	
231		361	380	399	418	436	455	474	493	511	530
232		549	568	586	605	624	642	661	680	698	717
233		736	754	773	791	810	829	847	866	884	903
234		922	940	959	977	996	*014	*033	*051	*070	*088
235	37 107	125	144	162	181	199	218	236	254	273	
236		291	310	328	346	365	383	401	420	438	457
237		475	493	511	530	548	566	585	603	621	639
238		658	676	694	712	731	749	767	785	803	822
239		840	858	876	894	912	931	949	967	985	*003
240	38 021	039	057	075	093	112	130	148	166	184	
241		202	220	238	256	274	292	310	328	346	364
242		382	399	417	435	453	471	489	507	525	543
243		561	578	596	614	632	650	668	686	703	721
244		739	757	775	792	810	828	846	863	881	899
245		917	934	952	970	987	*005	*023	*041	*058	*076
246	39 094	111	129	146	164	182	199	217	235	252	
247		270	287	305	322	340	358	375	393	410	428
248		445	463	480	498	515	533	550	568	585	602
249		620	637	655	672	690	707	724	742	759	777
250		794	811	829	846	863	881	898	915	933	950
N.	L. 0	1	2	3	4	5	6	7	8	9	P. P.

P. P.

22		21	
1	2,2		2,1
2	4,4		4,2
3	6,6		6,3
4	8,8		8,4
5	11,0		10,5
6	13,2		12,6
7	15,4		14,7
8	17,6		16,8
9	19,8		18,9

20	
1	2,0
2	4,0
3	6,0
4	8,0
5	10,0
6	12,0
7	14,0
8	16,0
9	18,0

19	
1	1,9
2	3,8
3	5,7
4	7,6
5	9,5
6	11,4
7	13,3
8	15,2
9	17,1

18	
1	1,8
2	3,6
3	5,4
4	7,2
5	9,0
6	10,8
7	12,6
8	14,4
9	16,2

17	
1	1,7
2	3,4
3	5,1
4	6,8
5	8,5
6	10,2
7	11,9
8	13,6
9	15,3

N.	L. 0	1	2	3	4	5	6	7	8	9
250	39 794	811	829	846	863	881	898	915	933	950
251	967	985	*002	*019	*037	*054	*071	*088	*106	*123
252	40 140	157	175	192	209	226	243	261	278	295
253	312	329	346	364	381	398	415	432	449	466
254	483	500	518	535	552	569	586	603	620	637
255	654	671	688	705	722	739	756	773	790	807
256	824	841	858	875	892	909	926	943	960	976
257	993	*010	*027	*044	*061	*078	*095	*111	*128	*145
258	41 162	179	196	212	229	246	263	280	296	313
259	330	347	363	380	397	414	430	447	464	481
260	497	514	531	547	564	581	597	614	631	647
261	664	681	697	714	731	747	764	.780	797	814
262	830	847	863	880	896	913	929	946	963	979
263	996	*012	*029	*045	*062	*078	*095	*111	*127	*144
264	42 160	177	193	210	226	243	259	275	292	308
265	325	341	357	374	390	406	423	439	455	472
266	488	504	521	537	553	570	586	602	619	635
267	651	667	684	700	716	732	749	765	781	797
268	813	830	846	862	878	894	911	927	943	959
269	975	991	*008	*024	*040	*056	*072	*088	*104	*120
270	43 136	152	169	185	201	217	233	249	265	281
271	297	313	329	345	361	377	393	409	425	441
272	457	473	489	505	521	537	553	569	584	600
273	616	632	648	664	680	696	712	727	743	759
274	775	791	807	823	838	854	870	886	902	917
275	933	949	965	981	996	*012	*028	*044	*059	*075
276	44 091	107	122	138	154	170	185	201	217	232
277	248	264	279	295	311	326	342	358	373	389
278	404	420	436	451	467	483	498	514	529	545
279	560	576	592	607	623	638	654	669	685	700
280	716	731	747	762	778	793	809	824	840	855
281	871	886	902	917	932	948	963	979	994	*010
282	45 025	040	056	071	086	102	117	133	148	163
283	179	194	209	225	240	255	271	286	301	317
284	332	347	362	378	393	408	423	439	454	469
285	484	500	515	530	545	561	576	591	606	621
286	637	652	667	682	697	712	728	743	758	773
287	788	803	818	834	849	864	879	894	909	924
288	939	954	969	984	*000	*015	*030	*045	*060	*075
289	46 090	105	120	135	150	165	180	195	210	225
290	240	255	270	285	300	315	330	345	359	374
291	389	404	419	434	449	464	479	494	509	523
292	538	553	568	583	598	613	627	642	657	672
293	687	702	716	731	746	761	776	790	805	820
294	835	850	864	879	894	909	923	938	953	967
295	982	997	*012	*026	*041	*056	*070	*085	*100	*114
296	47 129	144	159	173	188	202	217	232	246	261
297	276	290	305	319	334	349	363	378	392	407
298	422	436	451	465	480	494	509	524	538	553
299	567	582	596	611	625	640	654	669	683	698
300	712	727	741	756	770	784	799	813	828	842
N.	L. 0	1	2	3	4	5	6	7	8	9

P. P.

18

1	1,8
2	3,6
3	5,4
4	7,2
5	9,0
6	10,8
7	12,6
8	14,4
9	16,2

17

1	1,7
2	3,4
3	5,1
4	6,8
5	8,5
6	10,2
7	11,9
8	13,6
9	15,3

16

1	1,6
2	3,2
3	4,8
4	6,4
5	8,0
6	9,6
7	11,2
8	12,8
9	14,4

15

1	1,5
2	3,0
3	4,5
4	6,0
5	7,5
6	9,0
7	10,5
8	12,0
9	13,5

14

1	1,4
2	2,8
3	4,2
4	5,6
5	7,0
6	8,4
7	9,8
8	11,2
9	12,6

N.	L. 0	1	2	3	4	5	6	7	8	9	P. P.
300	47 712	727	741	756	770	784	799	813	828	842	
301	857	871	885	900	914	929	943	958	972	986	
302	48 001	015	029	044	058	073	087	101	116	130	
303	144	159	173	187	202	216	230	244	259	273	**15**
304	287	302	316	330	344	359	373	387	401	416	1 | 1,5
305	430	444	458	473	487	501	515	530	544	558	2 | 3,0
306	572	586	601	615	629	643	657	671	686	700	3 | 4,5 / 4 | 6,0
307	714	728	742	756	770	785	799	813	827	841	5 | 7,5
308	855	869	883	897	911	926	940	954	968	982	6 | 9,0
309	996	*010	*024	*038	*052	*066	*080	*094	*108	*122	7 | 10,5 / 8 | 12,0
310	49 186	150	164	178	192	206	220	234	248	262	9 | 13,5
311	276	290	304	318	332	346	360	374	388	402	
312	415	429	443	457	471	485	499	513	527	541	
313	554	568	582	596	610	624	638	651	665	679	
314	693	707	721	734	748	762	776	790	803	817	
315	831	845	859	872	886	900	914	927	941	955	**14**
316	969	982	996	*010	*024	*037	*051	*065	*079	*092	1 | 1,4
317	50 106	120	133	147	161	174	188	202	215	229	2 | 2,8
318	243	256	270	284	297	311	325	338	352	365	3 | 4,2 / 4 | 5,6
319	379	393	406	420	433	447	461	474	488	501	5 | 7,0
320	515	529	542	556	569	583	596	610	623	637	6 | 8,4 / 7 | 9,8
321	651	664	678	691	705	718	732	745	759	772	8 | 11,2
322	786	799	813	826	840	853	866	880	893	907	9 | 12,6
323	920	934	947	961	974	987	*001	*014	*028	*041	
324	51 055	068	081	095	108	121	135	148	162	175	
325	188	202	215	228	242	255	268	282	295	308	
326	322	335	348	362	375	388	402	415	428	441	
327	455	468	481	495	508	521	534	548	561	574	**13**
328	587	601	614	627	640	654	667	680	693	706	1 | 1,3
329	720	733	746	759	772	786	799	812	825	838	2 | 2,6
330	851	865	878	891	904	917	930	943	957	970	3 | 3,9 / 4 | 5,2
331	983	996	*009	*022	*035	*048	*061	*075	*088	*101	5 | 6,5
332	52 114	127	140	153	166	179	192	205	218	231	6 | 7,8 / 7 | 9,1
333	244	257	270	284	297	310	323	336	349	362	8 | 10,4
334	375	388	401	414	427	440	453	466	479	492	9 | 11,7
335	504	517	530	543	556	569	582	595	608	621	
336	634	647	660	673	686	699	711	724	737	750	
337	763	776	789	802	815	827	840	853	866	879	
338	892	905	917	930	943	956	969	982	994	*007	**12**
339	53 020	033	046	058	071	084	097	110	122	135	1 | 1,2
340	148	161	173	186	199	212	224	237	250	263	2 | 2,4
341	275	288	301	314	326	339	352	364	377	390	3 | 3,6 / 4 | 4,8
342	403	415	428	441	453	466	479	491	504	517	5 | 6,0
343	529	542	555	567	580	593	605	618	631	643	6 | 7,2 / 7 | 8,4
344	656	668	681	694	706	719	732	744	757	769	8 | 9,6
345	782	794	807	820	832	845	857	870	882	895	9 | 10,8
346	908	920	933	945	958	970	983	995	*008	*020	
347	54 033	045	058	070	083	095	108	120	133	145	
348	158	170	183	195	208	220	233	245	258	270	
349	283	295	307	320	332	345	357	370	382	394	
350	407	419	432	444	456	469	481	494	506	518	
N.	L. 0	1	2	3	4	5	6	7	8	9	P. P.

N.	L. 0	1	2	3	4	5	6	7	8	9	P. P.
350	54 407	419	432	444	456	469	481	494	506	518	
351	531	543	555	568	580	593	605	617	630	642	
352	654	667	679	691	704	716	728	741	753	765	
353	777	790	802	814	827	839	851	864	876	888	**18**
354	900	913	925	937	949	962	974	986	998	∗011	1 | 1,8
355	55 023	035	047	060	072	084	096	108	121	133	2 | 2,6
356	145	157	169	182	194	206	218	230	242	255	3 | 8,9
357	267	279	291	303	315	328	340	352	364	376	4 | 5,2
358	388	400	413	425	437	449	461	473	485	497	5 | 6,5
359	509	522	534	546	558	570	582	594	606	618	6 | 7,8
360	630	642	654	666	678	691	703	715	727	739	7 | 9,1
361	751	763	775	787	799	811	823	835	847	859	8 | 10,4
362	871	883	895	907	919	931	943	955	967	979	9 | 11,7
363	991	∗003	∗015	∗027	∗038	∗050	∗062	∗074	∗086	∗098	
364	56 110	122	134	146	158	170	182	194	205	217	
365	229	241	253	265	277	289	301	312	324	336	**12**
366	348	360	372	384	396	407	419	431	443	455	1 | 1,2
367	467	478	490	502	514	526	538	549	561	573	2 | 2,4
368	585	597	608	620	632	644	656	667	679	691	3 | 3,6
369	703	714	726	738	750	761	773	785	797	808	4 | 4,8
370	820	832	844	855	867	879	891	902	914	926	5 | 6,0
371	937	949	961	972	984	996	∗008	∗019	∗031	∗043	6 | 7,2
372	57 054	066	078	089	101	118	124	136	148	159	7 | 8,4
373	171	183	194	206	217	229	241	252	264	276	8 | 9,6
374	287	299	310	322	334	345	357	368	380	392	9 | 10,8
375	403	415	426	438	449	461	473	484	496	507	
376	519	530	542	553	565	576	588	600	611	623	
377	634	646	657	669	680	692	703	715	726	738	**11**
378	749	761	772	784	795	807	818	830	841	852	1 | 1,1
379	864	875	887	898	910	921	933	944	955	967	2 | 2,2
380	978	990	∗001	∗013	∗024	∗035	∗047	∗058	∗070	∗081	3 | 8,8
381	58 092	104	115	127	138	149	161	172	184	195	4 | 4,4
382	206	218	229	240	252	263	274	286	297	309	5 | 5,5
383	320	331	343	354	365	377	388	399	410	422	6 | 6,6
384	433	444	456	467	478	490	501	512	524	535	7 | 7,7
385	546	557	569	580	591	602	614	625	636	647	8 | 8,8
386	659	670	681	692	704	715	726	737	749	760	9 | 9,9
387	771	782	794	805	816	827	838	850	861	872	
388	883	894	906	917	928	939	950	961	973	984	
389	995	∗006	∗017	∗028	∗040	∗051	∗062	∗073	∗084	∗095	**10**
390	59 106	118	129	140	151	162	173	184	195	207	1 | 1,0
391	218	229	240	251	262	273	284	295	306	318	2 | 2,0
392	329	340	351	362	373	384	395	406	417	428	3 | 8,0
393	439	450	461	472	483	494	506	517	528	539	4 | 4,0
394	550	561	572	583	594	605	616	627	638	649	5 | 5,0
395	660	671	682	693	704	715	726	737	748	759	6 | 6,0
396	770	780	791	802	813	824	835	846	857	868	7 | 7,0
397	879	890	901	912	923	934	945	956	966	977	8 | 8,0
398	988	999	∗010	∗021	∗032	∗043	∗054	∗065	∗076	∗086	9 | 9,0
399	60 097	108	119	130	141	152	163	173	184	195	
400	206	217	228	239	249	260	271	282	293	304	
N.	L. 0	1	2	3	4	5	6	7	8	9	P. P.

L.	0	1	2	3	4	5	6	7	8	9		
	206	217	228	239	249	260	271	282	293	304		
	314	325	336	347	358	369	379	390	401	412		
	423	433	444	455	466	477	487	498	509	520		
	531	541	552	563	574	584	595	606	617	627		
	638	649	660	670	681	692	703	713	724	735		**11**
	746	756	767	778	788	799	810	821	831	842		
	853	863	874	885	895	906	917	927	938	949		1 | 1,1
	959	970	981	991	*002	*013	*023	*034	*045	*055		2 | 2,2
61	066	077	087	098	109	119	130	140	151	162		3 | 3,3
	172	183	194	204	215	225	236	247	257	268		4 | 4,4
	278	289	300	310	321	331	342	352	363	374		5 | 5,5
	384	395	405	416	426	437	448	458	469	479		6 | 6,6
	490	500	511	521	532	542	553	563	574	584		7 | 7,7
	595	606	616	627	637	648	658	669	679	690		8 | 8,8
	700	711	721	731	742	752	763	773	784	794		9 | 9,9
	805	815	826	836	847	857	868	878	888	899		
	909	920	930	941	951	962	972	982	993	*003		
62	014	024	034	045	055	066	076	086	097	107		
	118	128	138	149	159	170	180	190	201	211		
	221	232	242	252	263	273	284	294	304	315		
	325	335	346	356	366	377	387	397	408	418		
	428	439	449	459	469	480	490	500	511	521		**10**
	531	542	552	562	572	583	593	603	613	624		
	684	644	655	665	675	685	696	706	716	726		1 | 1,0
	737	747	757	767	778	788	798	808	818	829		2 | 2,0
	839	849	859	870	880	890	900	910	921	931		3 | 3,0
	941	951	961	972	982	992	*002	*012	*022	*033		4 | 4,0
												5 | 5,0
63	043	053	063	073	083	094	104	114	124	134		6 | 6,0
	144	155	165	175	185	195	205	215	225	236		7 | 7,0
	246	256	266	276	286	296	306	317	327	337		8 | 8,0
	347	357	367	377	387	397	407	417	428	438		9 | 9,0
	448	458	468	478	488	498	508	518	528	538		
	548	558	568	579	589	599	609	619	629	639		
	649	659	669	679	689	699	709	719	729	739		
	749	759	769	779	789	799	809	819	829	839		
	849	859	869	879	889	899	909	919	929	939		
	949	959	969	979	988	998	*008	*018	*028	*038		
64	048	058	068	078	088	098	108	118	128	137		**9**
	147	157	167	177	187	197	207	217	227	237		
	246	256	266	276	286	296	306	316	326	335		1 | 0,9
	345	355	365	375	385	395	404	414	424	434		2 | 1,8
	444	454	464	474	488	493	503	513	523	532		3 | 2,7
	542	552	562	572	582	591	601	611	621	631		4 | 3,6
	640	650	660	670	680	689	699	709	719	729		5 | 4,5
	738	748	758	768	777	787	797	807	816	826		6 | 5,4
	836	846	856	865	875	885	895	904	914	924		7 | 6,3
	933	943	953	963	972	982	992	*002	*011	*021		8 | 7,2
												9 | 8,1
65	031	040	050	060	070	079	089	099	108	118		
	128	137	147	157	167	176	186	196	205	215		
	225	234	244	254	263	273	283	292	302	312		
	321	331	341	350	360	369	379	389	398	408		

L.	0	1	2	3	4	5	6	7	8	9	P. P.

N.	L. 0	1	2	3	4	5	6	7	8	9	P. P.
450	65 821	831	841	850	860	369	379	389	398	408	
451	418	427	437	447	456	466	475	485	495	504	
452	514	523	533	543	552	562	571	581	591	600	
453	610	619	629	639	648	658	667	677	686	696	
454	706	715	725	734	744	753	763	772	782	792	**10**
455	801	811	820	830	839	849	858	868	877	887	1 1,0
456	896	906	916	925	935	944	954	963	973	982	2 2,0
457	992	*001	*011	*020	*030	*039	*049	*058	*068	*077	3 3,0
458	66 087	096	106	115	124	134	143	153	162	172	4 4,0
459	181	191	200	210	219	229	238	247	257	266	5 5,0
460	276	285	295	304	314	323	332	342	351	361	6 6,0
461	370	380	389	398	408	417	427	436	445	455	7 7,0
462	464	474	483	492	502	511	521	530	539	549	8 8,0
463	558	567	577	586	596	605	614	624	633	642	9 9,0
464	652	661	671	680	689	699	708	717	727	736	
465	745	755	764	773	783	792	801	811	820	829	
466	839	848	857	867	876	885	894	904	913	922	
467	932	941	950	960	969	978	987	997	*006	*015	
468	67 025	034	043	052	062	071	080	089	099	108	
469	117	127	136	145	154	164	173	182	191	201	
470	210	219	228	237	247	256	265	274	284	293	
471	302	311	321	330	339	348	357	367	376	385	**9**
472	394	403	413	422	431	440	449	459	468	477	1 0,9
473	486	495	504	514	523	532	541	550	560	569	2 1,8
474	578	587	596	605	614	624	633	642	651	660	3 2,7
475	669	679	688	697	706	715	724	733	742	752	4 3,6
476	761	770	779	788	797	806	815	825	834	843	5 4,5
477	852	861	870	879	888	897	906	916	925	934	6 5,4
478	943	952	961	970	979	988	997	*006	*015	*024	7 6,3
479	68 034	043	052	061	070	079	088	097	106	115	8 7,2
480	124	133	142	151	160	169	178	187	196	205	9 8,1
481	215	224	233	242	251	260	269	278	287	296	
482	305	314	323	332	341	350	359	368	377	386	
483	395	404	413	422	431	440	449	458	467	476	
484	485	494	503	511	520	529	538	547	556	565	
485	574	583	592	601	610	619	628	637	646	655	
486	664	673	681	690	699	708	717	726	735	744	
487	753	762	771	780	789	797	806	815	824	833	**8**
488	842	851	860	869	878	886	895	904	913	922	1 0,8
489	931	940	949	958	966	975	984	993	*002	*011	2 1,6
490	69 020	028	037	046	055	064	073	082	090	099	3 2,4
491	108	117	126	135	144	152	161	170	179	188	4 3,2
492	197	205	214	223	232	241	249	258	267	276	5 4,0
493	285	294	302	311	320	329	338	346	355	364	6 4,8
494	373	381	390	399	408	417	425	434	443	452	7 5,6
495	461	469	478	487	496	504	513	522	531	539	8 6,4
496	548	557	566	574	583	592	601	609	618	627	9 7,2
497	636	644	653	662	671	679	688	697	705	714	
498	723	732	740	749	758	767	775	784	793	801	
499	810	819	827	836	845	854	862	871	880	888	
500	897	906	914	923	932	940	949	958	966	975	
N.	L. 0	1	2	3	4	5	6	7	8	9	P. P.

71 014

N.	L. 0	1	2	3	4	5	6	7	8	9	P. P.
500	69 897	906	914	923	932	940	949	958	966	975	
501	984	992	*001	*010	*018	*027	*036	*044	*053	*062	
502	70 070	079	088	096	105	114	122	131	140	148	
503	157	165	174	183	191	200	209	217	226	234	
504	243	252	260	269	278	286	295	303	312	321	
505	329	338	346	355	364	372	381	389	398	406	
506	415	424	432	441	449	458	467	475	484	492	
507	501	509	518	526	535	544	552	561	569	578	
508	586	595	603	612	621	629	638	646	655	663	
509	672	680	689	697	706	714	723	731	740	749	
510	757	766	774	783	791	800	808	817	825	834	
511	842	851	859	868	876	885	893	902	910	919	
512	927	935	944	952	961	969	978	986	995	*003	
513	71 012	020	029	037	046	054	063	071	079	088	
514	096	105	113	122	130	139	147	155	164	172	
515	181	189	198	206	214	223	231	240	248	257	
516	265	273	282	290	299	307	315	324	332	341	
517	349	357	366	374	383	391	399	408	416	425	
518	433	441	450	458	466	475	483	492	500	508	
519	517	525	533	542	550	559	567	575	584	592	
520	600	609	617	625	634	642	650	659	667	675	
521	684	692	700	709	717	725	734	742	750	759	
522	767	775	784	792	800	809	817	825	834	842	
523	850	858	867	875	883	892	900	908	917	925	
524	933	941	950	958	966	975	983	991	999	*008	
525	72 016	024	032	041	049	057	066	074	082	090	
526	099	107	115	123	132	140	148	156	165	173	
527	181	189	198	206	214	222	230	239	247	255	
528	263	272	280	288	296	304	313	321	329	337	
529	346	354	362	370	378	387	395	403	411	419	
530	428	436	444	452	460	469	477	485	493	501	
531	509	518	526	534	542	550	558	567	575	583	
532	591	599	607	616	624	632	640	648	656	665	
533	673	681	689	697	705	713	722	730	738	746	
534	754	762	770	779	787	795	803	811	819	827	
535	835	843	852	860	868	876	884	892	900	908	
536	916	925	933	941	949	957	965	973	981	989	
537	997	*006	*014	*022	*030	*038	*046	*054	*062	*070	
538	73 078	086	094	102	111	119	127	135	143	151	
539	159	167	175	183	191	199	207	215	223	231	
540	239	247	255	263	272	280	288	296	304	312	
541	320	328	336	344	352	360	368	376	384	392	
542	400	408	416	424	432	440	448	456	464	472	
543	480	488	496	504	512	520	528	536	544	552	
544	560	568	576	584	592	600	608	616	624	632	
545	640	648	656	664	672	679	687	695	703	711	
546	719	727	735	743	751	759	767	775	783	791	
547	799	807	815	823	830	838	846	854	862	870	
548	878	886	894	902	910	918	926	933	941	949	
549	957	965	973	981	989	997	*005	*013	*020	*028	
550	74 036	044	052	060	068	076	084	092	099	107	
N.	L. 0	1	2	3	4	5	6	7	8	9	P. P.

P. P.

9
1 | 0,9
2 | 1,8
3 | 2,7
4 | 3,6
5 | 4,5
6 | 5,4
7 | 6,3
8 | 7,2
9 | 8,1

8
1 | 0,8
2 | 1,6
3 | 2,4
4 | 3,2
5 | 4,0
6 | 4,8
7 | 5,6
8 | 6,4
9 | 7,2

7
1 | 0,7
2 | 1,4
3 | 2,1
4 | 2,8
5 | 3,5
6 | 4,2
7 | 4,9
8 | 5,6
9 | 6,3

N.	L. 0	1	2	3	4	5	6	7	8	9	P. P.
550	74 086	044	052	060	068	076	084	092	099	107	
551	115	123	131	139	147	155	162	170	178	186	
552	194	202	210	218	225	233	241	249	257	265	
553	273	280	288	296	304	312	320	327	335	343	
554	351	359	367	374	382	390	398	406	414	421	
555	429	437	445	453	461	468	476	484	492	500	
556	507	515	523	531	539	547	554	562	570	578	
557	586	593	601	609	617	624	632	640	648	656	
558	663	671	679	687	695	703	710	718	726	733	
559	741	749	757	764	772	780	788	796	803	811	
560	819	827	834	842	850	858	865	873	881	889	
561	896	904	912	920	927	935	943	950	958	966	**8**
562	974	981	989	997	*005	*012	*020	*028	*035	*043	1 | 0,8
563	75 051	059	066	074	082	089	097	105	113	120	2 | 1,6
564	128	136	143	151	159	166	174	182	189	197	3 | 2,4
565	205	213	220	228	236	243	251	259	266	274	4 | 3,2
566	282	289	297	305	312	320	328	335	343	351	5 | 4,0
											6 | 4,8
567	358	366	374	381	389	397	404	412	420	427	7 | 5,6
568	435	442	450	458	465	473	481	488	496	504	8 | 6,4
569	511	519	526	534	542	549	557	565	572	580	9 | 7,2
570	587	595	603	610	618	626	633	641	648	656	
571	664	671	679	686	694	702	709	717	724	732	
572	740	747	755	762	770	778	785	793	800	808	
573	815	823	831	838	846	853	861	868	876	884	
574	891	899	906	914	921	929	937	944	952	959	
575	967	974	982	989	997	*005	*012	*020	*027	*035	
576	76 042	050	057	065	072	080	087	095	103	110	
577	118	125	133	140	148	155	163	170	178	185	
578	193	200	208	215	223	230	238	245	253	260	
579	268	275	283	290	298	305	313	320	328	335	
580	343	350	358	365	373	380	388	395	403	410	
581	418	425	433	440	448	455	462	470	477	485	**7**
582	492	500	507	515	522	530	537	545	552	559	1 | 0,7
583	567	574	582	589	597	604	612	619	626	634	2 | 1,4
584	641	649	656	664	671	678	686	693	701	708	3 | 2,1
585	716	723	730	738	745	753	760	768	775	782	4 | 2,8
586	790	797	805	812	819	827	834	842	849	856	5 | 3,5
											6 | 4,2
587	864	871	879	886	893	901	908	916	923	930	7 | 4,9
588	938	945	953	960	967	975	982	989	997	*004	8 | 5,6
589	77 012	019	026	034	041	048	056	063	070	078	9 | 6,3
590	085	093	100	107	115	122	129	137	144	151	
591	159	166	173	181	188	195	203	210	217	225	
592	232	240	247	254	262	269	276	283	291	298	
593	305	313	320	327	335	342	349	357	364	371	
594	379	386	393	401	408	415	422	430	437	444	
595	452	459	466	474	481	488	495	503	510	517	
596	525	532	539	546	554	561	568	576	583	590	
597	597	605	612	619	627	634	641	648	656	663	
598	670	677	685	692	699	706	714	721	728	735	
599	743	750	757	764	772	779	786	793	801	808	
600	815	822	830	837	844	851	859	866	873	880	
N.	L. 0	1	2	3	4	5	6	7	8	9	P. P.

N.	L. 0	1	2	3	4	5	6	7	8	9
600	77 815	822	830	837	844	851	859	866	873	880
601	887	895	902	909	916	924	931	938	945	952
602	960	967	974	981	988	996	*003	*010	*017	*025
603	78 032	039	046	053	061	068	075	082	089	097
604	104	111	118	125	132	140	147	154	161	168
605	176	183	190	197	204	211	219	226	233	240
606	247	254	262	269	276	283	290	297	305	312
607	319	326	333	340	347	355	362	369	376	383
608	390	398	405	412	419	426	433	440	447	455
609	462	469	476	483	490	497	504	512	519	526
610	533	540	547	554	561	569	576	583	590	597
611	604	611	618	625	633	640	647	654	661	668
612	675	682	689	696	704	711	718	725	732	739
613	746	753	760	767	774	781	789	796	803	810
614	817	824	831	838	845	852	859	866	873	880
615	888	895	902	909	916	923	930	937	944	951
616	958	965	972	979	986	993	*000	*007	*014	*021
617	79 029	036	043	050	057	064	071	078	085	092
618	099	106	113	120	127	134	141	148	155	162
619	169	176	183	190	197	204	211	218	225	232
620	239	246	253	260	267	274	281	288	295	302
621	309	316	323	330	337	344	351	358	365	372
622	379	386	393	400	407	414	421	428	435	442
623	449	456	463	470	477	484	491	498	505	511
624	518	525	532	539	546	553	560	567	574	581
625	588	595	602	609	616	623	630	637	644	650
626	657	664	671	678	685	692	699	706	713	720
627	727	734	741	748	754	761	768	775	782	789
628	796	803	810	817	824	831	837	844	851	858
629	865	872	879	886	893	900	906	913	920	927
630	934	941	948	955	962	969	975	982	989	996
631	80 003	010	017	024	030	037	044	051	058	065
632	072	079	085	092	099	106	113	120	127	134
633	140	147	154	161	168	175	182	188	195	202
634	209	216	223	229	236	243	250	257	264	271
635	277	284	291	298	305	312	318	325	332	339
636	346	353	359	366	373	380	387	393	400	407
637	414	421	428	434	441	448	455	462	468	475
638	482	489	496	502	509	516	523	530	536	543
639	550	557	564	570	577	584	591	598	604	611
640	618	625	632	638	645	652	659	665	672	679
641	686	693	699	706	713	720	726	733	740	747
642	754	760	767	774	781	787	794	801	808	814
643	821	828	835	841	848	855	862	868	875	882
644	889	895	902	909	916	922	929	936	943	949
645	956	963	969	976	983	990	996	*003	*010	*017
646	81 023	030	037	043	050	057	064	070	077	084
647	090	097	104	111	117	124	131	137	144	151
648	158	164	171	178	184	191	198	204	211	218
649	224	231	238	245	251	258	265	271	278	285
650	291	298	305	311	318	325	331	338	345	351
N.	L. 0	1	2	3	4	5	6	7	8	9

P. P.

8
1	0,8
2	1,6
3	2,4
4	3,2
5	4,0
6	4,8
7	5,6
8	6,4
9	7,2

7
1	0,7
2	1,4
3	2,1
4	2,8
5	3,5
6	4,2
7	4,9
8	5,6
9	6,3

6
1	0,6
2	1,2
3	1,8
4	2,4
5	3,0
6	3,6
7	4,2
8	4,8
9	5,4

N.	L. 0	1	2	3	4	5	6	7	8	9	P. P.
650	81 291	298	305	311	318	325	331	338	345	351	
651	358	365	371	378	385	391	398	405	411	418	
652	425	431	438	445	451	458	465	471	478	485	
653	491	498	505	511	518	525	531	538	544	551	
654	558	564	571	578	584	591	598	604	611	617	
655	624	631	637	644	651	657	664	671	677	684	
656	690	697	704	710	717	723	730	737	743	750	
657	757	763	770	776	783	790	796	803	809	816	
658	823	829	836	842	849	856	862	869	875	882	
659	889	895	902	908	915	921	928	935	941	948	
660	954	961	968	974	981	987	994	₊000	₊007	₊014	
661	82 020	027	033	040	046	053	060	066	073	079	
662	086	092	099	105	112	119	125	132	138	145	
663	151	158	164	171	178	184	191	197	204	210	
664	217	223	230	236	243	249	256	263	269	276	
665	282	289	295	302	308	315	321	328	334	341	
666	347	354	360	367	373	380	387	393	400	406	
667	413	419	426	432	439	445	452	458	465	471	
668	478	484	491	497	504	510	517	523	530	536	
669	543	549	556	562	569	575	582	588	595	601	
670	607	614	620	627	633	640	646	653	659	666	
671	672	679	685	692	698	705	711	718	724	730	
672	737	743	750	756	763	769	776	782	789	795	
673	802	808	814	821	827	834	840	847	853	860	
674	866	872	879	885	892	898	905	911	918	924	
675	930	937	943	950	956	963	969	975	982	988	
676	995	₊001	₊008	₊014	₊020	₊027	₊033	₊040	₊046	₊052	
677	83 059	065	072	078	085	091	097	104	110	117	
678	123	129	136	142	149	155	161	168	174	181	
679	187	193	200	206	213	219	225	232	238	245	
680	251	257	264	270	276	283	289	296	302	308	
681	315	321	327	334	340	347	353	359	366	372	
682	378	385	391	398	404	410	417	423	429	436	
683	442	448	455	461	467	474	480	487	493	499	
684	506	512	518	525	531	537	544	550	556	563	
685	569	575	582	588	594	601	607	613	620	626	
686	632	639	645	651	658	664	670	677	683	689	
687	696	702	708	715	721	727	734	740	746	753	
688	759	765	771	778	784	790	797	803	809	816	
689	822	828	835	841	847	853	860	866	872	879	
690	885	891	897	904	910	916	923	929	935	942	
691	948	954	960	967	973	979	985	992	998	₊004	
692	84 011	017	023	029	036	042	048	055	061	067	
693	073	080	086	092	098	105	111	117	123	130	
694	136	142	148	155	161	167	173	180	186	192	
695	198	205	211	217	223	230	236	242	248	255	
696	261	267	273	280	286	292	298	305	311	317	
697	323	330	336	342	348	354	361	367	373	379	
698	386	392	398	404	410	417	423	429	435	442	
699	448	454	460	466	473	479	485	491	497	504	
700	510	516	522	528	535	541	547	553	559	566	
N.	L. 0	1	2	3	4	5	6	7	8	9	P. P.

P. P.

7

1	0,7
2	1,4
3	2,1
4	2,8
5	3,5
6	4,2
7	4,9
8	5,6
9	6,3

6

1	0,6
2	1,2
3	1,8
4	2,4
5	3,0
6	3,6
7	4,2
8	4,8
9	5,4

N.	L. 0	1	2	3	4	5	6	7	8	9
700	84 510,	516	522	528	535	541	547	553	559	566
701	572	578	584	590	597	603	609	615	621	628
702	634	640	646	652	658	665	671	677	683	689
703	696	702	708	714	720	726	733	739	745	751
704	757	763	770	776	782	788	794	800	807	813
705	819	825	831	837	844	850	856	862	868	874
706	880	887	893	899	905	911	917	924	930	936
707	942	948	954	960	967	973	979	985	991	997
708	85 003	009	016	022	028	034	040	046	052	058
709	065	071	077	083	089	095	101	107	114	120
710	126	132	138	144	150	156	163	169	175	181
711	187	193	199	205	211	217	224	230	236	242
712	248	254	260	266	272	278	285	291	297	303
713	309	315	321	327	333	339	345	352	358	364
714	370	376	382	388	394	400	406	412	418	425
715	431	437	443	449	455	461	467	473	479	485
716	491	497	503	509	516	522	528	534	540	546
717	552	558	564	570	576	582	588	594	600	606
718	612	618	625	631	637	643	649	655	661	667
719	673	679	685	691	697	703	709	715	721	727
720	733	739	745	751	757	763	769	775	781	788
721	794	800	806	812	818	824	830	836	842	848
722	854	860	866	872	878	884	890	896	902	908
723	914	920	926	932	938	944	950	956	962	968
724	974	980	986	992	998	*004	*010	*016	*022	*028
725	86 034	040	046	052	058	064	070	076	082	088
726	094	100	106	112	118	124	130	136	141	147
727	153	159	165	171	177	183	189	195	201	207
728	213	219	225	231	237	243	249	255	261	267
729	273	279	285	291	297	303	308	314	320	326
730	332	338	344	350	356	362	368	374	380	386
731	392	398	404	410	415	421	427	433	439	445
732	451	457	463	469	475	481	487	493	499	504
733	510	516	522	528	534	540	546	552	558	564
734	570	576	581	587	593	599	605	611	617	623
735	629	635	641	646	652	658	664	670	676	682
736	688	694	700	705	711	717	723	729	735	741
737	747	753	759	764	770	776	782	788	794	800
738	806	812	817	823	829	835	841	847	853	859
739	864	870	876	882	888	894	900	906	911	917
740	923	929	935	941	947	953	958	964	970	976
741	982	988	994	999	*005	*011	*017	*023	*029	*035
742	87 040	046	052	058	064	070	075	081	087	093
743	099	105	111	116	122	128	134	140	146	151
744	157	163	169	175	181	186	192	198	204	210
745	216	221	227	233	239	245	251	256	262	268
746	274	280	286	291	297	303	309	315	320	326
747	332	338	344	349	355	361	367	373	379	384
748	390	396	402	408	413	419	425	431	437	442
749	448	454	460	466	471	477	483	489	495	500
750	506	512	518	523	529	535	541	547	552	558
N.	L. 0	1	2	3	4	5	6	7	8	9

P. P.

7

1	0,7
2	1,4
3	2,1
4	2,8
5	3,5
6	4,2
7	4,9
8	5,6
9	6,3

6

1	0,6
2	1,2
3	1,8
4	2,4
5	3,0
6	3,6
7	4,2
8	4,8
9	5,4

5

1	0,5
2	1,0
3	1,5
4	2,0
5	2,5
6	3,0
7	3,5
8	4,0
9	4,5

N.	L. 0	1	2	3	4	5	6	7	8	9	P. P.
750	87 506	512	518	523	529	535	541	547	552	558	
751	564	570	576	581	587	593	599	604	610	616	
752	622	628	633	639	645	651	656	662	668	674	
753	679	685	691	697	703	708	714	720	726	731	
754	737	743	749	754	760	766	772	777	783	789	
755	795	800	806	812	818	823	829	835	841	846	
756	852	858	864	869	875	881	887	892	898	904	
757	910	915	921	927	933	938	944	950	955	961	
758	967	973	978	984	990	996	*001	*007	*012	*018	
759	88 024	030	036	041	047	052	058	064	070	076	
760	081	087	093	098	104	110	116	121	127	133	
761	138	144	150	156	161	167	173	178	184	190	**6**
762	195	201	207	213	218	224	230	235	241	247	1 0,6
763	252	258	264	270	275	281	287	293	298	304	2 1,2
764	309	315	321	326	332	338	343	349	355	360	3 1,8
765	366	372	377	383	389	395	400	406	412	417	4 2,4
766	423	429	434	440	446	451	457	463	468	474	5 3,0
767	480	485	491	497	502	508	513	519	525	530	6 3,6
768	536	542	547	553	559	564	570	576	581	587	7 4,2
769	593	598	604	610	615	621	627	632	638	643	8 4,8
770	649	655	660	666	672	677	683	689	694	700	9 5,4
771	705	711	717	722	728	734	739	745	750	756	
772	762	767	773	779	784	790	795	801	807	812	
773	818	824	829	835	840	846	852	857	863	868	
774	874	880	885	891	897	902	908	913	919	925	
775	930	936	941	947	953	958	964	969	975	981	
776	986	992	997	*003	*009	*014	*020	*025	*031	*037	
777	89 042	048	053	059	064	070	076	081	087	092	
778	098	104	109	115	120	126	131	137	143	148	
779	154	159	165	170	176	182	187	193	198	204	
780	209	215	221	226	232	237	243	248	254	260	
781	265	271	276	282	287	293	298	304	310	315	**5**
782	321	326	332	337	343	348	354	360	365	371	1 0,5
783	376	382	387	393	398	404	409	415	421	426	2 1,0
784	432	437	443	448	454	459	465	470	476	481	3 1,5
785	487	492	498	504	509	515	520	526	531	537	4 2,0
786	542	548	553	559	564	570	575	581	586	592	5 2,5
787	597	603	609	614	620	625	631	636	642	647	6 3,0
788	653	658	664	669	675	680	686	691	697	702	7 3,5
789	708	713	719	724	730	735	741	746	752	757	8 4,0
790	763	768	774	779	785	790	796	801	807	812	9 4,5
791	818	823	829	834	840	845	851	856	862	867	
792	873	878	883	889	894	900	905	911	916	922	
793	927	933	938	944	949	955	960	966	971	977	
794	982	988	993	998	*004	*009	*015	*020	*026	*031	
795	90 037	042	048	053	059	064	069	075	080	086	
796	091	097	102	108	113	119	124	129	135	140	
797	146	151	157	162	168	173	179	184	189	195	
798	200	206	211	217	222	227	233	238	244	249	
799	255	260	266	271	276	282	287	293	298	304	
800	309	314	320	325	331	336	342	347	352	358	
N.	L. 0	1	2	3	4	5	6	7	8	9	P. P.

N.	L. 0	1	2	3	4	5	6	7	8	9	P. P.
800	90 309	314	320	325	331	336	342	347	352	358	
801	363	369	374	380	385	390	396	401	407	412	
802	417	423	428	434	439	445	450	455	461	466	
803	472	477	482	488	493	499	504	509	515	520	
804	526	531	536	542	547	552	558	563	569	574	
805	580	585	590	596	601	607	612	617	623	628	
806	634	639	644	650	655	660	666	671	677	682	
807	687	693	698	703	709	714	720	725	730	736	
808	741	747	752	757	763	768	773	779	784	789	
809	795	800	806	811	816	822	827	832	838	843	
810	849	854	859	865	870	875	881	886	891	897	
811	902	907	913	918	924	929	934	940	945	950	**6**
812	956	961	966	972	977	982	988	993	998	*004	
813	91 009	014	020	025	030	036	041	046	052	057	1 | 0,6
814	062	068	073	078	084	089	094	100	105	110	2 | 1,2
815	116	121	126	132	137	142	148	153	158	164	3 | 1,8
816	169	174	180	185	190	196	201	206	212	217	4 | 2,4
817	222	228	233	238	243	249	254	259	265	270	5 | 3,0
818	275	281	286	291	297	302	307	312	318	323	6 | 3,6
819	328	334	339	344	350	355	360	365	371	376	7 | 4,2
820	381	387	392	397	403	408	413	418	424	429	8 | 4,8
821	434	440	445	450	455	461	466	471	477	482	9 | 5,4
822	487	492	498	503	508	514	519	524	529	535	
823	540	545	551	556	561	566	572	577	582	587	
824	593	598	603	609	614	619	624	630	635	640	
825	645	651	656	661	666	672	677	682	687	693	
826	698	703	709	714	719	724	730	735	740	745	
827	751	756	761	766	772	777	782	787	793	798	
828	803	808	814	819	824	829	834	840	845	850	
829	855	861	866	871	876	882	887	892	897	903	
830	908	913	918	924	929	934	939	944	950	955	
831	960	965	971	976	981	986	991	997	*002	*007	**5**
832	92 012	018	023	028	033	038	044	049	054	059	
833	065	070	075	080	085	091	096	101	106	111	1 | 0,5
834	117	122	127	132	137	143	148	153	158	163	2 | 1,0
835	169	174	179	184	189	195	200	205	210	215	3 | 1,5
836	221	226	231	236	241	247	252	257	262	267	4 | 2,0
837	273	278	283	288	293	298	304	309	314	319	5 | 2,5
838	324	330	335	340	345	350	355	361	366	371	6 | 3,0
839	376	381	387	392	397	402	407	412	418	423	7 | 3,5
840	428	433	438	443	449	454	459	464	469	474	8 | 4,0
841	480	485	490	495	500	505	511	516	521	526	9 | 4,5
842	531	536	542	547	552	557	562	567	572	578	
843	583	588	593	598	603	609	614	619	624	629	
844	634	639	645	650	655	660	665	670	675	681	
845	686	691	696	701	706	711	716	722	727	732	
846	737	742	747	752	758	763	768	773	778	783	
847	788	793	799	804	809	814	819	824	829	834	
848	840	845	850	855	860	865	870	875	881	886	
849	891	896	901	906	911	916	921	927	932	937	
850	942	947	952	957	962	967	973	978	983	988	
N.	L. 0	1	2	3	4	5	6	7	8	9	P. P.

N.	L. 0	1	2	8	4	5	6	7	8	9	P. P.
850	92 942	947	952	957	962	967	973	978	983	988	
851	998	998	ₐ008	ₐ008	ₐ018	ₐ018	ₐ024	ₐ029	ₐ084	ₐ089	
852	93 044	049	054	059	064	069	075	080	085	090	
853	095	100	105	110	115	120	125	131	136	141	
854	146	151	156	161	166	171	176	181	186	192	
855	197	202	207	212	217	222	227	232	237	242	
856	247	252	258	263	268	273	278	283	288	293	
857	298	303	308	313	318	323	328	334	339	344	
858	349	354	359	364	369	374	379	384	389	394	**6**
859	399	404	409	414	420	425	430	435	440	445	1 │ 0,6
860	450	455	460	465	470	475	480	485	490	495	2 │ 1,2
861	500	505	510	515	520	526	531	536	541	546	3 │ 1,8
862	551	556	561	566	571	576	581	586	591	596	4 │ 2,4
863	601	606	611	616	621	626	631	636	641	646	5 │ 3,0
864	651	656	661	666	671	676	682	687	692	697	6 │ 8,6
865	· 702	707	712	717	722	727	732	737	742	747	7 │ 4,2
866	752	757	762	767	772	777	782	787	792	797	8 │ 4,8
867	802	807	812	817	822	827	832	837	842	847	9 │ 5,4
868	852	857	862	867	872	877	882	887	892	897	
869	902	907	912	917	922	927	932	937	942	947	
870	952	957	962	967	972	977	982	987	992	997	
871	94 002	007	012	017	022	027	032	037	042	047	
872	052	057	062	067	072	077	082	086	091	096	**5**
873	101	106	111	116	121	126	131	136	141	146	1 │ 0,5
874	151	156	161	166	171	176	181	186	191	196	2 │ 1,0
875	201	206	211	216	221	.226	231	236	240	245	3 │ 1,5
876	250	255	260	265	270	275	280	285	290	295	4 │ 2,0
877	300	305	310	315	320	325	330	335	340	345	5 │ 2,5
878	349	354	359	364	369	374	379	384	389	394	6 │ 3,0
879	399	404	409	414	419	424	429	433	438	443	7 │ 8,5
880	448	453	458	463	468	473	478	483	488	493	8 │ 4,0
881	498	503	507	512	517	522	527	532	537	542	9 │ 4,5
882	547	552	557	562	567	571	576	581	586	591	
883	596	601	606	611	616	621	626	630	635	640	
884	645	650	655	660	665	670	675	680	685	689	
885	694	699	704	709	714	719	724	729	734	738	
886	743	748	753	758	763	768	773	778	783	787	
887	792	797	802	807	812	817	822	827	832	836	**4**
888	841	846	851	856	861	866	871	876	880	885	1 │ 0,4
889	890	895	900	905	910	915	919	924	929	934	2 │ 0,8
890	939	944	949	954	959	963	968	973	978	983	3 │ 1,2
891	988	993	998	ₐ002	ₐ007	ₐ012	ₐ017	ₐ022	ₐ027	ₐ032	4 │ 1,6
892	95 036	041	046	051	056	061	066	071	075	080	5 │ 2,0
893	085	090	095	100	105	109	114	119	124	129	6 │ 2,4
894	134	139	143	148	153	158	163	168	173	177	7 │ 2,8
895	182	187	192	197	202	207	211	216	221	226	8 │ 3,2
896	231	236	240	245	250	255	260	265	270	274	9 │ 3,6
897	279	284	289	294	299	303	308	313	318	323	
898	328	332	337	342	347	352	357	361	366	371	
899	376	381	386	390	395	400	405	410	415	419	
900	424	429	434	439	444	448	453	458	463	468	
N.	L. 0	1	2	8	4	5	6	7	8	9	P. P.

N.	L. 0	1	2	3	4	5	6	7	8	9	P. P.
900	95 424	429	434	439	444	448	453	458	463	468	
901	472	477	482	487	492	497	501	506	511	516	
902	521	525	530	535	540	545	550	554	559	564	
903	569	574	578	583	588	593	598	602	607	612	
904	617	622	626	631	636	641	646	650	655	660	
905	665	670	674	679	684	689	694	698	703	708	
906	713	718	722	727	732	737	742	746	751	756	
907	761	766	770	775	780	785	789	794	799	804	
908	809	813	818	823	828	832	837	842	847	852	
909	856	861	866	871	875	880	885	890	895	899	
910	904	909	914	918	923	928	933	938	942	947	
911	952	957	961	966	971	976	980	985	990	995	**5**
912	999	*004	*009	*014	*019	*023	*028	*033	*038	*042	1 | 0,5
913	96 047	052	057	061	066	071	076	080	085	090	2 | 1,0
914	095	099	104	109	114	118	123	128	133	137	3 | 1,5
915	142	147	152	156	161	166	171	175	180	185	4 | 2,0
916	190	194	199	204	209	213	218	223	227	232	5 | 2,5
917	237	242	246	251	256	261	265	270	275	280	6 | 3,0
918	284	289	294	298	303	308	313	317	322	327	7 | 3,5
919	332	336	341	346	350	355	360	365	369	374	8 | 4,0
920	379	384	388	393	398	402	407	412	417	421	9 | 4,5
921	426	431	435	440	445	450	454	459	464	468	
922	473	478	483	487	492	497	501	506	511	515	
923	520	525	530	534	539	544	548	553	558	562	
924	567	572	577	581	586	591	595	600	605	609	
925	614	619	624	628	633	638	642	647	652	656	
926	661	666	670	675	680	685	689	694	699	703	
927	708	713	717	722	727	731	736	741	745	750	
928	755	759	764	769	774	778	783	788	792	797	
929	802	806	811	816	820	825	830	834	839	844	
930	848	853	858	862	867	872	876	881	886	890	
931	895	900	904	909	914	918	923	928	932	937	**4**
932	942	946	951	956	960	965	970	974	979	984	1 | 0,4
933	988	993	997	*002	*007	*011	*016	*021	*025	*030	2 | 0,8
934	97 035	039	044	049	053	058	063	067	072	077	3 | 1,2
935	081	086	090	095	100	104	109	114	118	123	4 | 1,6
936	128	132	137	142	146	151	155	160	165	169	5 | 2,0
937	174	179	183	188	192	197	202	206	211	216	6 | 2,4
938	220	225	230	234	239	243	248	253	257	262	7 | 2,8
939	267	271	276	280	285	290	294	299	304	308	8 | 3,2
940	313	317	322	327	331	336	340	345	350	354	9 | 3,6
941	359	364	368	373	377	382	387	391	396	400	
942	405	410	414	419	424	428	433	437	442	447	
943	451	456	460	465	470	474	479	483	488	493	
944	497	502	506	511	516	520	525	529	534	539	
945	543	548	552	557	562	566	571	575	580	585	
946	589	594	598	603	607	612	617	621	626	630	
947	635	640	644	649	653	658	663	667	672	676	
948	681	685	690	695	699	704	708	713	717	722	
949	727	731	736	740	745	749	754	759	763	768	
950	772	777	782	786	791	795	800	804	809	813	
N.	L. 0	1	2	3	4	5	6	7	8	9	P. P.

N.	L. 0	1	2	3	4	5	6	7	8	9	P. P.
950	97 772	777	782	786	791	795	800	804	809	813	
951	818	823	827	832	836	841	845	850	855	859	
952	864	868	873	877	882	886	891	896	900	905	
953	909	914	918	923	928	932	937	941	946	950	
954	955	959	964	968	973	978	982	987	991	996	
955	96 000	005	009	014	019	023	028	032	037	041	
956	046	050	055	059	064	068	073	078	082	087	
957	091	096	100	105	109	114	118	123	127	132	
958	137	141	146	150	155	159	164	168	173	177	
959	182	186	191	195	200	204	209	214	218	223	
960	227	232	236	241	245	250	254	259	263	268	
961	272	277	281	286	290	295	299	304	308	313	5
962	318	322	327	331	336	340	345	349	354	358	
963	363	367	372	376	381	385	390	394	399	403	1 | 0,5
964	408	412	417	421	426	430	435	439	444	448	2 | 1,0
965	453	457	462	466	471	475	480	484	489	493	3 | 1,5
966	498	502	507	511	516	520	525	529	534	538	4 | 2,0
967	543	547	552	556	561	565	570	574	579	583	5 | 2,5
968	588	592	597	601	605	610	614	619	623	628	6 | 3,0
969	632	637	641	646	650	655	659	664	668	673	7 | 3,5
970	677	682	686	691	695	700	704	709	713	717	8 | 4,0
971	722	726	731	735	740	744	749	753	758	762	9 | 4,5
972	767	771	776	780	784	789	793	798	802	807	
973	811	816	820	825	829	834	838	843	847	851	
974	856	860	865	869	874	878	883	887	892	896	
975	900	905	909	914	918	923	927	932	936	941	
976	945	949	954	958	968	967	972	976	981	985	
977	989	994	998	*003	*007	*012	*016	*021	*025	*029	
978	99 034	038	043	047	052	056	061	065	069	074	
979	078	083	087	092	096	100	105	109	114	118	
980	123	127	131	136	140	145	149	154	158	162	
981	167	171	176	180	185	189	193	198	202	207	4
982	211	216	220	224	229	233	238	242	247	251	
983	255	260	264	269	273	277	282	286	291	295	1 | 0,4
984	300	304	308	313	317	322	326	330	335	339	2 | 0,8
985	344	348	352	357	361	366	370	374	379	383	3 | 1,2
986	388	392	396	401	405	410	414	419	423	427	4 | 1,6
987	432	436	441	445	449	454	458	463	467	471	5 | 2,0
988	476	480	484	489	493	498	502	506	511	515	6 | 2,4
989	520	524	528	533	537	542	546	550	555	559	7 | 2,8
990	564	568	572	577	581	585	590	594	599	603	8 | 3,2
991	607	612	616	621	625	629	634	638	642	647	9 | 3,6
992	651	656	660	664	669	673	677	682	686	691	
993	695	699	704	708	712	717	721	726	730	734	
994	739	743	747	752	756	760	765	769	774	778	
995	782	787	791	795	800	804	808	813	817	822	
996	826	830	835	839	843	848	852	856	861	865	
997	870	874	878	883	887	891	896	900	904	909	
998	913	917	922	926	930	935	939	944	948	952	
999	957	961	965	970	974	978	983	987	991	996	
1000	00 000	004	009	013	017	022	026	030	035	039	
N.	L. 0	1	2	3	4	5	6	7	8	9	P. P.

	NUMBER.	LOGARITHM.
Base of Naperian logarithms	$e = 2.71828183$	0.4342945 .
Modulus of common logarithms	$u = 0.43429448$	9.6377843–10
Circumference of a circle in degrees . .	$= 360$	2.5563025
Circumference of a circle in minutes . .	$= 21600$	4.3344538
Circumference of a circle in seconds . .	$= 1296000$	6.1126050
Radius of a circle in degrees	$= 57.29578$	1.7581226
Radius of a circle in minutes	$= 3437.7468$	3.5362739
Radius of a circle in seconds	$= 206264.806$	5.3144251
Ratio of a circumference to diameter . .	$\pi = 3.14159265$	0.4971499

NUMBER.	LOGARITHM.		NUMBER.	LOGARITHM.
$2\pi = 6.28318531$	0.7981799		$\pi^2 = 9.86960440$	0.9942997
$4\pi = 12.56637061$	1.0992099		$\dfrac{1}{\pi} = 0.10132118$	9.0057003–10
$\dfrac{\pi}{2} = 1.57079633$	0.1961199		$\sqrt{\pi} = 1.77245385$	0.2485749
$\dfrac{\pi}{3} = 1.04719755$	0.0200286		$\dfrac{1}{\sqrt{\pi}} = 0.56418958$	9.7514251–10
$\dfrac{4\pi}{3} = 4.18879020$	0.6220886		$\sqrt{\dfrac{3}{\pi}} = 0.97720502$	9.9899857–10
$\dfrac{\pi}{4} = 0.78539816$	9.8950899–10		$\sqrt{\dfrac{4}{\pi}} = 1.12837917$	0.0524551
$\dfrac{\pi}{6} = 0.52359878$	9.7189986–10		$\sqrt[3]{\pi} = 1.46459189$	0.1657166
$\dfrac{1}{\pi} = 0.31830989$	9.5028501–10		$\dfrac{1}{\sqrt[3]{\pi}} = 0.68278406$	9.8342834–10
$\dfrac{1}{2\pi} = 0.15915494$	9.2018201–10		$\sqrt[3]{\pi^2} = 2.14502940$	0.3314332
$\dfrac{3}{\pi} = 0.95492966$	9.9799714–10		$\sqrt[3]{\dfrac{3}{4\pi}} = 0.62035049$	9.7926371–10
$\dfrac{4}{\pi} = 1.27323954$	0.1049101		$\sqrt[3]{\dfrac{\pi}{6}} = 0.80599598$	9.9063329–10
$\dfrac{3}{4\pi} = 0.23873241$	9.3779114–10			

If the radius $r = 1$, the length of the arc is:

			LOGARITHM.
for 1 degree $=$	$\dfrac{\pi}{180}$	$= 0.01745329$	8.2418774–10
for 1 minute $=$	$\dfrac{\pi}{10800}$	$= 0.00029089$	6.4637261–10
for 1 second $=$	$\dfrac{\pi}{648000}$	$= 0.00000485$	4.6855749–10
for $\frac{1}{2}$ degree $=$	$\dfrac{\pi}{360}$	$= 0.00872665$	7.9408474–10
for $\frac{1}{2}$ minute $=$	$\dfrac{\pi}{21600}$	$= 0.00014544$	6.1626961–10
for $\frac{1}{2}$ second $=$	$\dfrac{\pi}{1296000}$	$= 0.00000242$	4.3845449–10
sin $1''$ in the unit circle $= 0.00000485$			4.6855749–10

TABLE II.

LOGARITHMS OF THE TRIGONOMETRIC FUNCTIONS,

From 0° to 1° and 89° to 90° for every second,

From 1° to 6° and 84° to 89° for every ten seconds,

From 6° to 84° for every minute.

L. Cos.   L. Sin.   **0°**   L. Tang.

0.00	' ''		0''	1''	2''	8''	4''	5''	6''	7''	8''	9''	10''	
000	0 0	4.	—	63557	98660	₊16270	₊28768	₊38454	₊46878	₊58067	₊58866	₊68982	₊68557	50
000	10	5.	68557	72697	76476	79952	83170	86167	88969	91602	94085	96488	98660	40
000	20		98660	₊00779	₊02800	₊04780	₊06579	₊08851	₊10055	₊11694	₊13278	₊14797	₊16270	80
000	80	6.	16270	17694	19072	20409	21705	22964	24188	25878	26586	27664	28768	20
000	40		28768	29886	80882	81904	82908	83879	84883	85767	86682	87577	88454	10
000	50		88454	89815	40158	40985	41797	42594	48876	44145	44900	45648	46878	0 59
000	1 0	6.4	6878	7090	7797	8492	9175	9849	₊0512	₊1165	₊1808	₊2442	₊8067	50
000	10	6.5	8067	8688	4291	4890	5481	6064	6689	7207	7767	8820	8866	40
000	20		8866	9406	9989	₊0465	₊0985	₊1499	₊2007	₊2509	₊8006	₊8496	₊8982	80
000	80	6.6	8982	4462	4986	5406	5870	6880	6785	7285	7680	8121	8557	20
000	40		8557	8990	9418	9841	₊0261	₊0676	₊1088	₊1496	₊1900	₊2800	₊2697	10
000	50	6.7	2697	8090	8479	8865	4248	4627	5008	5876	5746	6112	6476	0 58
000	2 0		6476	6886	7198	7548	7900	8248	8595	8988	9278	9616	9952	50
000	10		9952	₊0255	₊0615	₊0948	₊1268	₊1591	₊1911	₊2280	₊2545	₊2859	₊8170	40
000	20	6.8	8170	8479	8786	4091	4894	4694	4998	5289	5584	5876	6167	80
000	80		6167	6455	6742	7027	7810	7591	7870	8147	8428	8697	8969	20
000	40		5969	9240	9509	9776	₊0042	₊0806	₊0568	₊0829	₊1088	₊1846	₊1602	10
000	50	6.9	1602	1857	2110	2362	2612	2861	8109	8855	8599	8848	4085	0 57
000	8 0		4085	4825	4565	4808	5089	·5275	5509	5742	5978	6204	6488	50
000	10		6488	6661	6888	7118	7888	7561	7788	8004	8224	8448	8660	40
000	20		8660	8877	9098	9807	9520	9788	9944	₊0155	₊0864	₊0572	₊0779	80
000	80	7.0	0779	0986	1191	1895	1599	1801	2008	2208	2408	2602	2800	20
000	40		2800	2997	8198	8888	8582	8776	8968	4160	4851	4541	4780	10
000	50		4780	4919	5106	5298	5479	5604	5849	6082	6215	6897	6579	0 56
000	4 0		6579	6759	6989	7118	7296	7474	7651	7827	8008	8177	8851	50
000	10		8851	8525	8698	8870	9041	9211	9881	9551	9719	9887	₊0055	40
000	20	7.1	0055	0222	0888	0558	0718	0882	1046	1209	1871	1588	1694	80
000	80		1694	1854	2014	2174	2888	2491	2648	2805	2962	8118	8278	20
000	40		8278	8428	8582	8786	8889	4042	4194	4846	4497	4647	4797	10
000	50		4797	4947	5096	5244	5892	5540	5687	5888	5979	6125	6270	0 55
0.00			10''	9''	8''	7''	6''	5''	4''	8''	2''	1''	0''	'' '

L. Sin.   L. Cos.   **89°**   L. Cotg.

	144	143	142	141	140	139		138	137	136	135	134	133	
1	14,4	14,3	14,2	14,1	14,0	13,9	1	13,8	13,7	13,6	13,5	13,4	13,3	1
2	28,8	28,6	28,4	28,2	28,0	27,8	2	27,6	27,4	27,2	27,0	26,8	26,6	2
3	43,2	42,9	42,6	42,3	42,0	41,7	3	41,4	41,1	40,8	40,5	40,2	39,9	3
4	57,6	57,2	56,8	56,4	56,0	55,6	4	55,2	54,8	54,4	54,0	53,6	53,2	4
5	72,0	71,5	71,0	70,5	70,0	69,5	5	69,0	68,5	68,0	67,5	67,0	66,5	5
6	86,4	85,8	85,2	84,6	84,0	83,4	6	82,8	82,2	81,6	81,0	80,4	79,8	6
7	100,8	100,1	99,4	98,7	98,0	97,3	7	96,6	95,9	95,2	94,5	93,8	93,1	7
8	115,2	114,4	113,6	112,8	112,0	111,2	8	110,4	109,6	108,8	108,0	107,2	106,4	8
9	129,6	128,7	127,8	126,9	126,0	125,1	9	124,2	123,3	122,4	121,5	120,6	119,7	9

	132	131	130	129	128	127		126	125	124	123	122	121	
1	13,2	13,1	13,0	12,9	12,8	12,7	1	12,6	12,5	12,4	12,3	12,2	12,1	1
2	26,4	26,2	26,0	25,8	25,6	25,4	2	25,2	25,0	24,8	24,6	24,4	24,2	2
3	39,6	39,3	39,0	38,7	38,4	38,1	3	37,8	37,5	37,2	36,9	36,6	36,3	3
4	52,8	52,4	52,0	51,6	51,2	50,8	4	50,4	50,0	49,6	49,2	48,8	48,4	4
5	66,0	65,5	65,0	64,5	64,0	63,5	5	63,0	62,5	62,0	61,5	61,0	60,5	5
6	79,2	78,6	78,0	77,4	76,8	76,2	6	75,6	75,0	74,4	73,8	73,2	72,6	6
7	92,4	91,7	91,0	90,3	89,6	88,9	7	88,2	87,5	86,8	86,1	85,4	84,7	7
8	105,6	104,8	104,0	103,2	102,4	101,6	8	100,8	100,0	99,2	98,4	97,6	96,8	8
9	118,8	117,9	117,0	116,1	115,2	114,3	9	113,4	112,5	111,6	110,7	109,8	108,9	9

	120	119	118	117	116	115		114	113	112	111	110	109	
1	12,0	11,9	11,8	11,7	11,6	11,5	1	11,4	11,3	11,2	11,1	11,0	10,9	1
2	24,0	23,8	23,6	23,4	23,2	23,0	2	22,8	22,6	22,4	22,2	22,0	21,8	2
3	36,0	35,7	35,4	35,1	34,8	34,5	3	34,2	33,9	33,6	33,3	33,0	32,7	3
4	48,0	47,6	47,2	46,8	46,4	46,0	4	45,6	45,2	44,8	44,4	44,0	43,6	4
5	60,0	59,5	59,0	58,5	58,0	57,5	5	57,0	56,5	56,0	55,5	55,0	54,5	5
6	72,0	71,4	70,8	70,2	69,6	69,0	6	68,4	67,8	67,2	66,6	66,0	65,4	6
7	84,0	83,3	82,6	81,9	81,2	80,5	7	79,8	79,1	78,4	77,7	77,0	76,3	7
8	96,0	95,2	94,4	93,6	92,8	92,0	8	91,2	90,4	89,6	88,8	88,0	87,2	8
9	108,0	107,1	106,2	105,3	104,4	103,5	9	102,6	101,7	100,8	99,9	99,0	98,1	9

0.00	' "	0″	1″	2″	3″	4″	5″	6″	7″	8″	9″	10″	
000	5 0	7.1 6270	6414	6558	6702	6845	6987	7130	7271	7413	7558	7694	50
000	10	7694	7834	7973	8112	8250	8389	8526	8668	8800	8937	9072	40
000	20	9072	9208	9343	9478	9612	9746	9879	*0012	*0145	*0277	*0409	30
000	30	7.2 0409	0540	0671	0802	0932	1062	1191	1320	1449	1577	1705	20
000	40	1705	1833	1960	2087	2218	2339	2465	2590	2715	2840	2964	10
000	50	2964	3088	3212	3335	3458	3580	3702	3824	3946	4067	4188	0 54
000	6 0	4188	4308	4428	4548	4668	4787	4906	5024	5142	5260	5378	50
000	10	5378	5495	5612	5728	5845	5961	6076	6192	6307	6421	6536	40
000	20	6536	6650	6764	6877	6991	7104	7216	7329	7441	7552	7664	30
000	30	7664	7775	7886	7997	8107	8217	8327	8437	8546	8655	8763	20
000	40	8763	8872	8980	9088	9196	9308	9410	9517	9628	9736	9836	10
000	50	9836	9942	*0047	*0152	*0257	*0362	*0467	*0571	*0675	*0779	*0882	0 53
000	7 0	7.3 0682	0786	0890	1191	1294	1396	1498	1600	1702	1808	1904	50
000	10	1904	2005	2106	2206	2306	2406	2506	2606	2705	2804	2903	40
000	20	2903	3001	3100	3198	3296	3393	3491	3588	3685	3782	3879	30
000	30	3879	3975	4071	4167	4268	4363	4454	4549	4644	4739	4833	20
000	40	4833	4928	5022	5116	5209	5303	5396	5489	5582	5675	5767	10
000	50	5767	5860	5952	6044	6135	6227	6318	6409	6500	6591	6682	0 52
000	8 0	6682	6772	6862	6952	7042	7132	7221	7310	7399	7488	7577	50
000	10	7577	7666	7754	7842	7930	8018	8106	8193	8280	8367	8454	40
000	20	8454	8541	8628	8714	8800	8887	8972	9058	9144	9229	9314	30
000	30	9314	9400	9484	9569	9654	9738	9822	9906	9990	*0074	*0158	20
000	40	7.4 0158	0241	0324	0408	0491	0573	0656	0739	0821	0903	0985	10
000	50	0985	1067	1149	1230	1312	1393	1474	1555	1636	1716	1797	0 51
000	9 0	1797	1877	1957	2037	2117	2197	2277	2356	2435	2515	2594	50
000	10	2594	2673	2751	2830	2908	2987	3065	3143	3221	3299	3376	40
000	20	3376	3454	3531	3608	3685	3762	3839	3916	3992	4069	4145	30
000	30	4145	4221	4297	4373	4449	4524	4600	4675	4750	4825	4900	20
000	40	4900	4975	5050	5124	5199	5273	5347	5421	5495	5569	5643	10
000	50	5643	5716	5790	5863	5936	6009	6082	6155	6228	6300	6373	0 50
0.00		10″	9″	8″	7″	6″	5″	4″	3″	2″	1″	0″	″ '

L. Tang.   0°

	108	107	106	105	104	103		102	101	99	98	97	96	
1	10,8	10,7	10,6	10,5	10,4	10,3	1	10,2	10,1	9,9	9,8	9,7	9,6	1
2	21,6	21,4	21,2	21,0	20,8	20,6	2	20,4	20,2	19,8	19,6	19,4	19,2	2
3	32,4	32,1	31,8	31,5	31,2	30,9	3	30,6	30,3	29,7	29,4	29,1	28,8	3
4	43,2	42,8	42,4	42,0	41,6	41,2	4	40,8	40,4	39,6	39,2	38,8	38,4	4
5	54,0	53,5	53,0	52,5	52,0	51,5	5	51,0	50,5	49,5	49,0	48,5	48,0	5
6	64,8	64,2	63,6	63,0	62,4	61,8	6	61,2	60,6	59,4	58,8	58,2	57,6	6
7	75,6	74,9	74,2	73,5	72,8	72,1	7	71,4	70,7	69,3	68,6	67,9	67,2	7
8	86,4	85,6	84,8	84,0	83,2	82,4	8	81,6	80,8	79,2	78,4	77,6	76,8	8
9	97,2	96,3	95,4	94,5	93,6	92,7	9	91,8	90,9	89,1	88,2	87,3	86,4	9

	95	94	93	92	91	90		89	88	87	86	85	84	
1	9,5	9,4	9,3	9,2	9,1	9,0	1	8,9	8,8	8,7	8,6	8,5	8,4	1
2	19,0	18,8	18,6	18,4	18,2	18,0	2	17,8	17,6	17,4	17,2	17,0	16,8	2
3	28,5	28,2	27,9	27,6	27,3	27,0	3	26,7	26,4	26,1	25,8	25,5	25,2	3
4	38,0	37,6	37,2	36,8	36,4	36,0	4	35,6	35,2	34,8	34,4	34,0	33,6	4
5	47,5	47,0	46,5	46,0	45,5	45,0	5	44,5	44,0	43,5	43,0	42,5	42,0	5
6	57,0	56,4	55,8	55,2	54,6	54,0	6	53,4	52,8	52,2	51,6	51,0	50,4	6
7	66,5	65,8	65,1	64,4	63,7	63,0	7	62,3	61,6	60,9	60,2	59,5	58,8	7
8	76,0	75,2	74,4	73,6	72,8	72,0	8	71,2	70,4	69,6	68,8	68,0	67,2	8
9	85,5	84,6	83,7	82,8	81,9	81,0	9	80,1	79,2	78,3	77,4	76,5	75,6	9

	83	82	81	80	79	78		77	76	75	74	73	72	
1	8,3	8,2	8,1	8,0	7,9	7,8	1	7,7	7,6	7,5	7,4	7,3	7,2	1
2	16,6	16,4	16,2	16,0	15,8	15,6	2	15,4	15,2	15,0	14,8	14,6	14,4	2
3	24,9	24,6	24,3	24,0	23,7	23,4	3	23,1	22,8	22,5	22,2	21,9	21,6	3
4	33,2	32,8	32,4	32,0	31,6	31,2	4	30,8	30,4	30,0	29,6	29,2	28,8	4
5	41,5	41,0	40,5	40,0	39,5	39,0	5	38,5	38,0	37,5	37,0	36,5	36,0	5
6	49,8	49,2	48,6	48,0	47,4	46,8	6	46,2	45,6	45,0	44,4	43,8	43,2	6
7	58,1	57,4	56,7	56,0	55,3	54,6	7	53,9	53,2	52,5	51,8	51,1	50,4	7
8	66,4	65,6	64,8	64,0	63,2	62,4	8	61,6	60,8	60,0	59,2	58,4	57,6	8
9	74,7	73,8	72,9	72,0	71,1	70,2	9	69,3	68,4	67,5	66,6	65,7	64,8	9

′ ″	0″	1″	2″	3″	4″	5″	6″	7″	8″	9″	10″		
5 0	7.1 6270	6414	6558	6702	6845	6988	7180	7271	7413	7555	7694	50	
10	7694	7834	7973	8112	8250	8389	8526	8668	8800	8937	9073	40	
20	9073	9208	9343	9478	9612	9746	9879	*0012	*0145	*0277	*0409	30	
30	7.2 0409	0540	0671	0802	0932	1062	1191	1321	1449	1577	1705	20	
40	1705	1833	1960	2087	2218	2339	2465	2590	2715	2840	2964	10	
50	2964	3088	3212	3335	3458	3580	3703	3824	3946	4067	4188	0 54	
6 0	4188	4306	4428	4548	4668	4787	4906	5024	5142	5260	5378	50	
10	5378	5495	5612	5728	5845	5961	6076	6192	6307	6421	6586	40	
20	6586	6650	6764	6877	6991	7104	7216	7329	7441	7552	7664	30	
30	7664	7775	7886	7997	8107	8217	8327	8437	8546	8655	8764	20	
40	8764	8872	8980	9088	9196	9303	9410	9517	9624	9730	9836	10	
50	9836	9942	*0047	*0153	*0258	*0362	*0467	*0571	*0675	*0779	*0882	0 53	
7 0	7.3 0882	0986	1089	1192	1294	1396	1499	1600	1702	1803	1904	50	
10	1904	2005	2106	2206	2307	2406	2506	2606	2705	2804	2903	40	
20	2903	3001	3100	3198	3296	3394	3491	3588	3685	3782	3879	30	
30	3879	3975	4071	4167	4263	4359	4454	4549	4644	4739	4833	20	
40	4833	4928	5022	5116	5209	5303	5396	5489	5582	5675	5767	10	
50	5767	5860	5952	6044	6135	6227	6318	6409	6500	6591	6682	0 52	
8 0	6682	6772	6862	6952	7042	7182	7221	7310	7400	7488	7577	50	
10	7577	7666	7754	7842	7930	8018	8106	8193	8281	8368	8455	40	
20	8455	8541	8628	8714	8801	8887	8973	9058	9144	9229	9315	30	
30	9315	9400	9485	9569	9654	9738	9823	9907	9991	*0074	*0158	20	
40	7.4 0158	0241	0325	0408	0491	0574	0656	0739	0821	0903	0985	10	
50	0985	1067	1149	1230	1312	1393	1474	1555	1636	1716	1797	0 51	
9 0	1797	1877	1958	2038	2117	2197	2277	2356	2436	2515	2594	50	
10	2594	2673	2751	2830	2909	2987	3065	3143	3221	3299	3376	40	
20	3376	3454	3531	3608	3686	3762	3839	3916	3992	4069	4145	30	
30	4145	4221	4297	4373	4449	4524	4600	4675	4750	4825	4900	20	
40	4900	4975	5050	5124	5199	5273	5347	5421	5495	5569	5643	10	
50	5643	5716	5790	5863	5936	6009	6082	6155	6228	6300	6373	0 50	
	10″	9″	8″	7″	6″	5″	4″	3″	2″	1″	0″	″ ′	

L. Sin.   89°   L. Cotg.

L. Cos.   **L. Sin.**   **0°**

0.00	' ''	0''	1''	2''	3''	4''	5''	6''	7''	8''	9''	10''		d.	P. P.
000	10 0	7.46 373	445	517	589	661	733	805	876	948	*019	*090	50	72	**72**
000	10	7.47 090	162	233	303	374	445	515	586	656	*726	797	40	71	1 ǀ 7,2
000	20	797	867	936	*006	*076	*145	*215	*284	*353	*422	*491	30	69	2 ǀ 14,4
000	30	7.48 491	560	629	698	766	835	903	971	*039	*108	*175	20	68	3 ǀ 21,6
000	40	7.49 175	243	311	379	446	513	581	648	715	782	849	10	67	4 ǀ 28,8
000	50	849	916	982	*049	*115	*182	*248	*314	*380	*446	*512	0 49	66	5 ǀ 36,0
															6 ǀ 43,2
															7 ǀ 50,4
000	11 0	7.50 512	578	643	709	774	840	905	970	*035	*100	*165	50	65	8 ǀ 57,6
000	10	7.51 165	230	294	359	423	488	552	616	680	744	808	40	64	9 ǀ 64,8
000	20	808	872	936	999	*063	*126	*190	*253	*316	*379	*442	30	63	**70**
000	30	7.52 442	505	568	631	693	756	818	881	943	*005	*067	20	62	1 ǀ 7,0
000	40	7.53 067	129	191	253	315	376	438	499	561	622	683	10	62	2 ǀ 14,0
000	50	683	744	805	866	927	988	*049	*109	*170	*230	*291	0 48	61	3 ǀ 21,0
															4 ǀ 28,0
															5 ǀ 35,0
000	12 0	7.54 291	351	411	471	531	591	651	711	771	830	890	50	60	6 ǀ 42,0
000	10	890	949	*009	*068	*127	*186	*245	*304	*363	*422	*481	40	59	7 ǀ 49,0
000	20	7.55 481	539	598	656	715	773	831	889	948	*006	*064	30	58	8 ǀ 56,0
000	30	7.56 064	121	179	237	295	352	410	467	524	582	639	20	58	9 ǀ 63,0
000	40	639	696	753	810	867	924	980	*037	*094	*150	*206	10	57	**68**
000	50	7.57 206	263	319	375	431	488	544	599	655	711	767	0 47	56	1 ǀ 6,8
															2 ǀ 13,6
															3 ǀ 20,4
000	13 0	767	822	878	934	989	*044	*100	*155	*210	*265	*320	50	55	4 ǀ 27,2
000	10	7.58 320	375	430	485	539	594	649	703	758	812	866	40	55	5 ǀ 34,0
000	20	866	921	975	*029	*083	*137	*191	*245	*299	*352	*406	30	54	6 ǀ 40,8
000	30	7.59 406	459	513	566	620	673	726	780	833	886	939	20	53	7 ǀ 47,6
000	40	989	992	*045	*097	*150	*203	*255	*308	*360	*413	*465	10	53	8 ǀ 54,4
000	50	7.60 465	517	570	622	674	726	778	830	882	934	985	0 46	52	9 ǀ 61,2
															66
000	14 0	985	*037	*089	*140	*192	*243	*294	*346	*397	*448	*499	50	51	1 ǀ 6,6
000	10	7.61 499	550	601	652	703	754	805	855	906	957	*007	40	51	2 ǀ 13,2
000	20	7.62 007	058	108	158	209	259	309	359	409	459	509	30	50	3 ǀ 19,8
000	30	509	559	609	659	708	758	808	857	907	956	*006	20	50	4 ǀ 26,4
000	40	7.63 006	055	104	158	208	252	301	350	399	448	496	10	49	5 ǀ 33,0
000	50	496	545	594	642	691	740	788	837	885	933	982	0 45	49	6 ǀ 39,6
															7 ǀ 46,2
															8 ǀ 52,8
000	15 0	982	*030	*078	*126	*174	*222	*270	*318	*366	*414	*461	50	48	9 ǀ 59,4
000	10	7.64 461	509	557	604	652	699	747	794	842	889	936	40	48	**64**
000	20	936	983	*030	*078	*125	*172	*218	*265	*312	*359	*406	30	47	1 ǀ 6,4
000	30	7.65 406	452	499	546	592	638	685	731	778	824	870	20	46	2 ǀ 12,8
000	40	870	916	962	*009	*055	*101	*146	*192	*238	*284	*330	10	46	3 ǀ 19,2
000	50	7.66 330	375	421	467	512	558	603	649	694	789	784	0 44	45	4 ǀ 25,6
															5 ǀ 32,0
															6 ǀ 38,4
000	16 0	784	830	875	920	965	*010	*055	*100	*145	*190	*235	50	45	7 ǀ 44,8
000	10	7.67 235	279	324	369	413	458	502	547	591	686	680	40	44	8 ǀ 51,2
*000	20	680	724	768	818	857	901	945	989	*033	*077	*121	30	44	9 ǀ 57,6
*999	30	7.68 121	165	208	252	296	340	383	427	470	514	557	20	44	**62**
999	40	557	601	644	687	731	774	817	860	903	946	989	10	43	1 ǀ 6,2
999	50	989	*032	*075	*118	*161	*204	*247	*289	*332	*375	*417	0 43	43	2 ǀ 12,4
															3 ǀ 18,6
999	17 0	7.69 417	460	502	545	587	630	672	714	757	799	841	50	42	4 ǀ 24,8
999	10	841	883	925	967	*009	*051	*093	*135	*177	*219	*261	40	42	5 ǀ 31,0
999	20	7.70 261	302	344	386	427	469	510	552	593	635	676	30	42	6 ǀ 37,2
999	30	676	718	759	800	841	883	924	965	*006	*047	*088	20	41	7 ǀ 43,4
999	40	7.71 088	129	170	211	251	292	333	374	414	455	496	10	41	8 ǀ 49,6
999	50	496	536	577	617	658	698	739	779	819	859	900	0 42	40	9 ǀ 55,8
															61
999	18 0	900	940	980	*020	*060	*100	*140	*180	*220	*260	*300	50	40	1 ǀ 6,1
999	10	7.72 300	340	380	419	459	499	538	578	618	657	697	40	40	2 ǀ 12,2
999	20	697	736	775	815	854	894	933	972	*011	*050	*090	30	39	3 ǀ 18,3
999	30	7.73 090	129	168	207	246	285	324	363	401	440	479	20	39	4 ǀ 24,4
999	40	479	518	557	595	634	673	711	750	788	827	865	10	39	5 ǀ 30,5
999	50	865	904	942	980	*019	*057	*095	*133	*171	*210	*248	0 41	38	6 ǀ 36,6
															7 ǀ 42,7
															8 ǀ 48,8
999	19 0	7.74 248	286	324	362	400	438	476	514	551	589	627	50	38	9 ǀ 54,9
999	10	627	665	703	740	778	815	853	891	928	966	*003	40	38	**60**
999	20	7.75 003	040	078	115	153	190	227	264	302	339	376	30	37	1 ǀ 6,0
999	30	376	413	450	487	524	561	598	635	672	709	745	20	37	2 ǀ 12,0
999	40	745	782	819	856	892	929	966	*002	*039	*075	*112	10	37	3 ǀ 18,0
999	50	7.76 112	148	185	221	258	294	330	367	403	439	475	0 40	36	4 ǀ 24,0
															5 ǀ 30,0
															6 ǀ 36,0
															7 ǀ 42,0
															8 ǀ 48,0
9.99		10''	9''	8''	7''	6''	5''	4''	3''	2''	1''	0''	'' '	d.	9 ǀ 54,0
															P. P.

L. Sin.   **89°**   L. Cos.

′ ′	0″	1″	2″	8″	4″	5″	6″	7″	8″	9″	10″		d.	P. P.

	0″	1″	2″	8″	4″	5″	6″	7″	8″	9″	10″		d.	P. P.
10 0	7.46 373	445	517	589	661	733	805	876	948	*019	*091	50	72	**59 58 57**
10	7.47 091	162	288	804	874	445	516	586	656	727	797	40	71	1 5,9 5,8 5,7
20	797	867	987	*006	*076	*146	*215	*284	*854	*428	*492	30	70	2 11,8 11,6 11,4
30	7.48 492	561	629	698	767	885	903	972	*040	*108	*176	20	68	3 17,7 17,4 17,1
40	7.49 176	243	311	879	446	514	581	648	715	782	849	10	67	4 23,6 23,2 22,8
50	849	916	982	*049	*115	*182	*248	*314	*380	*446	*512	0 **49**	66	5 29,5 29,0 28,5
														6 35,4 34,8 34,2
														7 41,8 40,6 89,9
11 0	7.50 512	578	648	709	774	840	905	970	*085	*100	*165	50	65	8 47,2 46,4 45,6
10	7.51 165	230	295	859	424	488	552	617	681	745	809	40	64	9 53,1 52,2 51,3
20	809	872	986	*000	*063	*127	*190	*258	*316	*380	*443	30	68	**56 55 54**
30	7.52 443	505	568	631	694	756	819	881	948	*005	*067	20	62	1 5,6 5,5 5,4
40	7.53 067	129	191	258	315	377	488	500	561	622	688	10	62	2 11,2 11,0 10,8
50	688	745	806	867	927	988	*049	*110	*170	*231	*291	0 **48**	61	3 16,8 16,5 16,2
														4 22,4 22,0 21,6
														5 28,0 27,5 27,0
12 0	7.54 291	351	411	471	532	591	651	711	771	880	890	50	60	6 33,6 33,0 32,4
10	890	949	*009	*068	*127	*186	*245	*304	*868	*422	*481	40	59	7 39,2 38,5 37,8
20	7.55 481	539	598	657	715	773	832	890	948	*006	*064	30	58	8 44,8 44,0 43,2
30	7.56 064	122	179	237	295	352	410	467	525	582	639	20	58	9 50,4 49,5 48,6
40	639	696	758	810	867	924	981	*037	*094	*150	*207	10	57	**53 52 51**
50	7.57 207	268	819	376	432	488	544	600	656	711	767	0 **47**	56	1 5,3 5,2 5,1
														2 10,6 10,4 10,2
														3 15,9 15,6 15,3
18 0	767	823	878	934	989	*045	*100	*155	*210	*265	*320	50	55	4 21,2 20,8 20,4
10	7.58 320	875	480	485	540	594	649	704	758	812	867	40	55	5 26,5 26,0 25,5
20	867	921	975	*029	*088	*137	*191	*245	*299	*858	*406	30	54	6 31,8 31,2 30,6
30	7.59 406	460	518	567	620	673	727	780	888	886	989	20	53	7 37,1 36,4 35,7
40	989	992	*045	*098	*150	*203	*256	*808	*361	*413	*466	10	58	8 42,4 41,6 40,8
50	7.60 466	518	570	622	674	726	778	880	882	934	986	0 **46**	52	9 47,7 46,8 45,9
														50 49 48
														1 5,0 4,9 4,8
14 0	986	*037	*089	*140	*192	*243	*295	*346	*897	*449	*500	50	51	2 10,0 9,8 9,6
10	7.61 500	551	602	653	704	754	805	856	906	957	*008	40	51	3 15,0 14,7 14,4
20	7.62 008	058	108	159	209	259	810	860	410	460	510	30	50	4 20,0 19,6 19,2
30	510	560	609	659	709	759	808	858	907	957	*006	20	50	5 25,0 24,5 24,0
40	7.63 006	055	105	154	203	252	801	350	899	448	497	10	49	6 30,0 29,4 25,6
50	497	546	594	643	692	740	789	837	885	934	982	0 **45**	48	7 35,0 34,3 33,6
														8 40,0 39,2 38,4
														9 45,0 44,1 43,2
15 0	982	*080	*078	*127	*175	*228	*271	*318	*866	*414	*462	50	48	**47 46 45**
10	7.64 462	510	557	605	652	700	747	795	842	889	987	40	48	1 4,7 4,6 4,5
20	987	984	*031	*078	*125	*172	*219	*266	*318	*859	*406	30	47	2 9,4 9,2 9,0
30	7.65 406	458	499	546	592	639	685	732	778	824	871	20	46	3 14,1 13,8 13,5
40	871	917	963	*009	*055	*101	*147	*198	*289	*284	*380	10	46	4 18,8 18,4 18,0
50	7.66 380	376	421	467	513	558	604	649	694	740	785	0 **44**	46	5 23,5 23,0 22,5
														6 28,2 27,6 27,0
														7 32,9 32,2 31,5
16 0	785	830	875	920	966	*011	*056	*100	*145	*190	*235	50	45	8 37,6 36,8 36,0
10	7.67 235	280	824	869	414	458	508	547	592	636	680	40	44	9 42,3 41,4 40,5
20	680	725	769	813	857	901	946	990	*084	*077	*121	30	44	**44 43 42**
30	7.68 121	165	209	253	296	840	884	427	471	514	558	20	44	1 4,4 4,3 4,2
40	558	601	645	688	731	774	818	861	904	947	990	10	43	2 8,8 8,6 8,4
50	990	*033	*076	*119	*162	*204	*247	*290	*888	*875	*418	0 **48**	43	3 13,2 12,9 12,6
														4 17,6 17,2 16,8
														5 22,0 21,5 21,0
17 0	7.69 418	460	503	545	588	680	673	715	757	799	842	50	42	6 26,4 25,8 25,2
10	842	884	926	968	*010	*052	*094	*186	*178	*219	*261	40	42	7 30,8 30,1 29,4
20	7.70 261	303	845	886	428	469	511	558	594	685	677	30	42	8 35,2 34,4 33,6
30	677	718	759	801	842	883	924	965	*006	*047	*088	20	41	9 39,6 38,7 37,8
40	7.71 088	129	170	211	252	293	834	374	415	456	496	10	41	**41 40 39**
50	496	537	577	618	658	699	789	779	820	860	900	0 **42**	40	1 4,1 4,0 3,9
														2 8,2 8,0 7,8
														3 12,3 12,0 11,7
18 0	900	940	981	*021	*061	*101	*141	*181	*221	*261	*801	50	40	4 16,4 16,0 15,6
10	7.72 301	340	880	420	460	499	589	579	618	658	697	40	40	5 20,5 20,0 19,5
20	697	737	776	815	855	894	933	978	*012	*051	*090	30	39	6 24,6 24,0 23,4
30	7.73 090	129	168	207	246	285	824	863	402	441	480	20	39	7 28,7 28,0 27,8
40	480	518	557	596	685	678	712	750	789	827	866	10	39	8 32,8 32,0 31,2
50	866	904	948	981	*019	*058	*096	*184	*172	*210	*248	0 **41**	38	9 36,9 36,0 35,1
														38 37 36
														1 3,8 3,7 8,6
19 0	7.74 248	286	825	868	401	438	476	514	552	590	628	50	38	2 7,6 7,4 7,2
10	628	665	703	741	779	816	854	891	929	966	*004	40	38	3 11,4 11,1 10,8
20	7.75 004	041	079	116	153	191	228	265	302	889	877	30	37	4 15,2 14,8 14,4
30	877	414	451	488	525	562	599	686	672	709	746	20	37	5 19,0 18,5 18,0
40	746	788	820	856	898	930	966	*003	*040	*076	*118	10	37	6 22,8 22,2 21,6
50	7.76 113	149	186	222	258	295	881	867	404	440	476	0 **40**	36	7 26,6 25,9 25,2
														8 30,4 29,6 28,8
														9 34,2 33,8 32,4

	10″	9″	8″	7″	6″	5″	4″	8″	2″	1″	0″		d.	P. P.

L. Cos. **L. Sin.** **0°**

9.99	′ ″	0″	1″	2″	3″	4″	5″	6″	7″	8″	9″	10″		d.
999	**20** 0	7.76 475	512	548	584	620	656	692	728	764	800	886	50	36
999	10	886	872	907	948	979	*015	*051	*086	*122	*158	*193	40	36
999	20	7.77 198	229	264	300	3₃5	371	406	442	477	512	548	30	36
999	30	548	583	618	654	689	724	759	794	829	864	899	20	35
999	40	899	934	969	*004	*089	*074	*109	*144	*179	*218	*248	10	35
999	50	7.78 248	283	318	352	887	422	456	491	525	560	594	0 **39**	35
999	**21** 0	594	629	663	698	732	766	801	835	869	903	988	50	34
999	10	938	972	*006	*040	*074	*108	*142	*176	*210	*244	*278	40	34
999	20	7.79 273	312	346	380	414	448	481	515	549	582	616	30	34
999	30	616	650	683	717	751	784	818	851	885	918	952	20	34
999	40	952	985	*018	*052	*085	*118	*152	*185	*218	*251	*284	10	33
999	50	7.80 284	317	351	884	417	450	483	516	549	582	615	0 **38**	33
999	**22** 0	615	647	680	713	746	779	812	844	877	910	942	50	33
999	10	942	975	*008	*040	*073	*105	*188	*170	*203	*235	*268	40	33
999	20	7.81 268	300	332	365	397	429	462	494	526	558	591	30	32
999	30	591	623	655	687	719	751	788	815	847	879	911	20	32
999	40	911	943	975	*007	*089	*070	*102	*184	*166	*198	*229	10	32
999	50	7.82 229	261	293	324	356	387	419	451	482	514	545	0 **37**	32
999	**23** 0	545	577	608	639	671	702	783	765	796	827	859	50	31
999	10	859	890	921	952	988	*015	*046	*077	*108	*189	*170	40	31
999	20	7.83 170	201	232	263	294	325	356	387	417	448	479	30	31
999	30	479	510	541	571	602	633	663	694	725	755	786	20	31
999	40	786	817	847	878	908	939	969	*000	*080	*060	*091	10	30
999	50	7.84 091	121	151	182	212	242	273	303	333	363	393	0 **36**	30
999	**24** 0	393	424	454	484	514	544	574	604	634	664	694	50	30
999	10	694	724	754	784	814	843	873	903	933	963	992	40	30
999	20	992	*022	*052	*082	*111	*141	*171	*200	*280	*259	*289	30	30
999	30	7.85 289	318	348	377	407	436	466	495	525	554	583	20	29
999	40	588	613	642	671	701	780	759	788	817	847	876	10	29
999	50	876	905	984	963	992	*021	*050	*079	*108	*187	*166	0 **35**	29
999	**25** 0	7.86 166	195	224	258	282	311	340	368	897	426	455	50.	29
999	10	455	484	512	541	570	598	627	656	684	718	741	40	29
999	20	741	770	799	827	856	884	918	941	969	998	*026	30	28
999	30	7.87 026	055	083	111	140	168	196	224	258	281	809	20	28
999	40	309	337	366	394	422	450	478	506	584	562	590	10	28
999	50	590	618	646	674	702	780	758	786	814	842	870	0 **34**	28
999	**26** 0	870	897	925	958	981	*009	*086	*064	*092	*119	*147	50	28
999	10	7.88 147	175	202	230	258	285	318	340	368	395	428	40	28
999	20	428	450	478	505	533	560	587	615	642	669	697	30	27
999	30	697	724	751	779	806	833	860	888	915	942	969	20	27
999	40	969	996	*023	*050	*077	*105	*132	*159	*186	*218	*240	10	27
999	50	7.89 240	267	294	820	347	374	401	428	455	482	509	0 **33**	27
999	**27** 0	509	535	562	589	616	642	669	696	722	749	776	50	27
999	10	776	802	829	856	882	909	935	962	988	*015	*041	40	26
999	20	7.90 041	068	094	121	147	174	200	226	253	279	305	30	26
999	30	305	332	358	884	411	437	463	489	515	542	568	20	26
999	40	568	594	620	646	672	698	725	751	777	803	829	10	26
999	50	829	855	881	907	933	958	984	*010	*036	*062	*088	0 **32**	26
999	**28** 0	7.91 088	114	140	165	191	217	243	269	294	320	846	50	26
999	10	346	371	397	428	448	474	500	525	551	576	602	40	26
999	20	602	627	653	678	704	729	755	780	806	831	857	30	26
999	30	857	882	907	933	958	983	*009	*034	*059	*085	*110	20	25
998	40	7.92 110	135	160	186	211	236	261	286	311	336	362	10	25
998	50	362	387	412	437	462	487	512	537	562	587	612	0 **31**	25
998	**29** 0	612	637	662	687	712	787	761	786	811	886	861	50	25
998	10	861	886	910	935	960	985	*009	*034	*059	*084	*108	40	25
998	20	7.93 108	133	158	182	207	231	256	281	305	330	854	30	25
998	30	854	379	403	428	452	477	501	526	550	575	599	20	24
998	40	599	623	648	672	696	721	745	769	794	818	842	10	24
998	50	842	866	891	915	939	963	988	*012	*036	*060	*084	0 **30**	24
9.99		10″	9″	8″	7″	6″	5″	4″	3″	2″	1″	0″	″ ′	d.

L. Sin. **89°** **L. Cos.**

L. Tang.   0°

′ ″	0″	1″	2″	3″	4″	5″	6″	7″	8″	9″	10″		d.
20 0	7.76 476	512	548	585	621	657	698	729	765	801	837	50	36
10	837	872	908	944	980	*016	*051	*087	*123	*158	*194	40	36
20	7.77 194	230	265	301	336	372	407	442	478	513	549	30	36
30	549	584	619	654	690	725	760	795	830	865	900	20	35
40	900	935	970	*005	*040	*075	*110	*145	*179	*214	*249	10	35
50	7.78 249	284	318	353	388	422	457	492	526	561	595	0 39	35
21 0	595	630	664	698	733	767	801	836	870	904	938	50	34
10	938	973	*007	*041	*075	*109	*143	*177	*211	*245	*279	40	34
20	7.79 279	313	347	381	415	448	482	516	550	583	617	30	34
30	617	651	684	718	751	785	819	852	886	919	952	20	34
40	952	986	*019	*053	*086	*119	*152	*186	*219	*252	*285	10	33
50	7.80 285	318	351	385	418	451	484	517	550	583	615	0 38	33
22 0	615	648	681	714	747	780	812	845	878	911	943	50	33
10	943	976	*009	*041	*074	*106	*139	*171	*204	*236	*269	40	33
20	7.81 269	301	333	366	398	430	463	495	527	559	591	30	32
30	591	624	656	688	720	752	784	816	848	880	912	20	32
40	912	944	976	*008	*040	*071	*103	*135	*167	*198	*230	10	32
50	7.82 230	262	294	325	357	388	420	452	483	515	546	0 37	32
23 0	546	578	609	640	672	703	734	766	797	828	860	50	31
10	860	891	922	953	984	*016	*047	*078	*109	*140	*171	40	31
20	7.83 171	202	233	264	295	326	357	388	418	449	480	30	31
30	480	511	542	572	603	634	664	695	726	756	787	20	31
40	787	818	848	879	909	940	970	*001	*031	*061	*092	10	30
50	7.84 092	122	152	183	213	243	274	304	334	364	394	0 36	30
24 0	394	425	455	485	515	545	575	605	635	665	695	50	30
10	695	725	755	785	815	845	874	904	934	964	993	40	30
20	993	*023	*053	*083	*112	*142	*172	*201	*231	*260	*290	30	30
30	7.85 290	319	349	378	408	437	467	496	526	555	584	20	29
40	584	614	643	672	702	731	760	789	819	848	877	10	29
50	877	906	935	964	993	*022	*051	*080	*109	*138	*167	0 35	29
25 0	7.86 167	196	225	254	283	312	341	370	398	427	456	50	29
10	456	485	513	542	571	600	628	657	685	714	743	40	29
20	743	771	800	828	857	885	914	942	971	999	*027	30	28
30	7.87 027	056	084	113	141	169	197	226	254	282	310	20	28
40	310	339	367	395	423	451	479	507	535	563	591	10	28
50	591	619	647	675	703	731	759	787	815	843	871	0 34	28
26 0	871	899	926	954	982	*010	*037	*065	*093	*121	*148	50	28
10	7.88 148	176	204	231	259	286	314	342	369	397	424	40	28
20	424	452	479	506	534	561	589	616	643	671	698	30	27
30	698	725	753	780	807	834	862	889	916	943	970	20	27
40	970	997	*025	*052	*079	*106	*133	*160	*187	*214	*241	10	27
50	7.89 241	268	295	322	349	376	403	429	456	483	510	0 33	27
27 0	510	537	563	590	617	644	670	697	724	750	777	50	27
10	777	804	830	857	884	910	937	963	990	*016	*043	40	27
20	7.90 043	069	096	122	149	175	201	228	254	280	307	30	26
30	307	333	359	386	412	438	464	491	517	543	569	20	26
40	569	595	622	648	674	700	726	752	778	804	830	10	26
50	830	856	882	908	934	960	986	*012	*038	*064	*089	0 32	26
28 0	7.91 089	115	141	167	193	218	244	270	296	321	347	50	26
10	347	373	398	424	450	475	501	527	552	578	603	40	26
20	603	629	654	680	705	731	756	782	807	833	858	30	26
30	858	883	909	934	960	985	*010	*036	*061	*086	*111	20	25
40	7.92 111	137	162	187	212	237	263	288	313	338	363	10	25
50	363	388	413	438	463	488	513	538	563	588	613	0 31	25
29 0	613	638	663	688	713	738	763	788	813	838	862	50	25
10	862	887	912	937	961	986	*011	*036	*060	*085	*110	40	25
20	7.93 110	134	159	184	208	233	258	282	307	331	356	30	25
30	356	380	405	429	454	478	503	527	552	576	601	20	24
40	601	625	649	674	698	722	747	771	795	820	844	10	24
50	844	868	892	917	941	965	989	*013	*038	*062	*086	0 30	24
	10″	9″	8″	7″	6″	5″	4″	3″	2″	1″	0″	″ ′	d.

P. P.

	37	36
1	3.7	3.6
2	7.4	7.2
3	11.1	10.8
4	14.8	14.4
5	18.5	18.0
6	22.2	21.6
7	25.9	25.2
8	29.6	28.8
9	33.3	32.4

	35	34
1	3.5	3.4
2	7.0	6.8
3	10.5	10.2
4	14.0	13.6
5	17.5	17.0
6	21.0	20.4
7	24.5	23.8
8	28.0	27.2
9	31.5	30.6

	33	32
1	3.3	3.2
2	6.6	6.4
3	9.9	9.6
4	13.2	12.8
5	16.5	16.0
6	19.8	19.2
7	23.1	22.4
8	26.4	25.6
9	29.7	28.8

	31	30
1	3.1	3.0
2	6.2	6.0
3	9.3	9.0
4	12.4	12.0
5	15.5	15.0
6	18.6	18.0
7	21.7	21.0
8	24.8	24.0
9	27.9	27.0

	29	28
1	2.9	2.8
2	5.8	5.6
3	8.7	8.4
4	11.6	11.2
5	14.5	14.0
6	17.4	16.8
7	20.3	19.6
8	23.2	22.4
9	26.1	25.2

	27	26
1	2.7	2.6
2	5.4	5.2
3	8.1	7.8
4	10.8	10.4
5	13.5	13.0
6	16.2	15.6
7	18.9	18.2
8	21.6	20.8
9	24.3	23.4

	25	24
1	2.5	2.4
2	5.0	4.8
3	7.5	7.2
4	10.0	9.6
5	12.5	12.0
6	15.0	14.4
7	17.5	16.8
8	20.0	19.2
9	22.5	21.6

L. Cos.   **L. Sin.**   **0°**

9.99	' ''	0"	1"	2"	3"	4"	5"	6"	7"	8"	9"	10"	.	d.
998	**30** 0	7.94 064	108	132	157	.181	205	229	253	277	301	325	50	24
998	10	325	349	373	397	421	445	469	492	516	540	564	40	24
998	20	564	588	612	636	659	683	707	731	755	778	802	30	24
998	30	802	826	849	873	897	921	944	968	991	*015	*089	20	24
998	40	7.95 089	062	086	109	133	157	180	204	227	251	274	10	24
998	50	274	298	321	344	368	391	415	438	461	485	508	0 **29**	23
998	**31** 0	508	532	555	578	601	625	648	671	695	718	741	50	23
998	10	741	764	787	811	884	857	880	903	926	950	973	40	23
998	20	973	996	*019	*042	*065	*088	*111	*184	*157	*180	*203	30	23
998	30	7.96 203	226	249	272	295	318	341	364	386	409	432	20	23
998	40	432	455	478	501	524	546	569	592	615	637	660	10	23
998	50	660	683	706	728	751	774	796	819	842	864	887	0 **28**	23
998	**32** 0	887	910	932	955	977	*000	*022	*045	*068	*090	*118	50	23
998	10	7.97 118	135	158	180	202	225	247	270	292	315	337	40	22
998	20	337	359	382	404	426	449	471	493	516	538	560	30	22
998	30	560	583	605	627	649	672	694	716	738	760	782	20	22
998	40	782	804	827	849	871	898	915	937	959	981	*008	10	22
998	50	7.98 003	025	048	070	092	114	136	157	179	201	223	0 **27**	22
998	**33** 0	223	245	267	289	311	333	355	377	398	420	442	50	22
998	10	442	464	486	508	529	551	573	595	616	638	660	40	22
998	20	660	682	703	725	747	768	790	812	833	855	876	30	22
998	30	876	898	920	941	963	984	*006	*027	*049	*070	*092	20	22
998	40	7.99 092	113	135	156	178	199	221	242	264	285	306	10	21
998	50	306	328	349	371	392	413	435	456	477	499	520	0 **26**	21
998	**34** 0	520	541	562	584	605	626	647	669	690	711	732	50	21
998	10	732	753	775	796	817	838	859	880	901	922	943	40	21
998	20	943	965	986	*007	*028	*049	*070	*091	*112	*133	*154	30	21
998	30	8.00 154	175	196	217	238	259	279	300	321	342	363	20	21
998	40	363	384	405	426	447	467	488	509	530	551	571	10	21
998	50	571	592	613	634	654	675	696	717	737	758	779	0 **25**	21
998	**35** 0	779	799	820	841	861	882	903	923	944	964	985	50	21
998	10	985	*006	*026	*047	*067	*088	*108	*129	*149	*170	*190	40	20
998	20	8.01 190	211	231	252	272	293	313	333	354	374	395	30	20
998	30	395	415	435	456	476	496	517	537	557	578	598	20	20
998	40	598	618	639	659	679	699	720	740	760	801	*002	10	20
998	50	801	821	841	861	881	901	922	942	962	982	*002	0 **24**	20
998	**36** 0	8.02 002	022	042	062	082	102	123	143	163	183	203	50	20
998	10	203	223	243	263	283	303	323	343	362	382	402	40	20
998	20	402	422	442	462	482	502	522	542	561	581	601	30	20
998	30	601	621	641	661	680	700	720	740	759	779	799	20	20
998	40	799	819	838	858	878	898	917	937	957	976	996	10	20
998	50	996	*016	*035	*055	*074	*094	*114	*133	*153	*172	*192	0 **23**	20
997	**37** 0	8.03 192	212	231	251	270	290	309	329	348	368	387	50	20
997	10	387	407	426	446	465	484	504	523	543	562	581	40	19
997	20	581	601	620	640	659	678	698	717	736	756	775	30	19
997	30	775	794	813	833	852	871	891	910	929	948	967	20	19
997	40	967	987	*006	*025	*044	*063	*083	*102	*121	*140	*159	10	19
997	50	8.04 159	178	197	217	236	255	274	293	312	331	350	0 **22**	19
997	**38** 0	350	369	388	407	426	445	464	483	502	521	540	50	19
997	10	540	559	578	597	616	635	654	673	692	710	729	40	19
997	20	729	748	767	786	805	824	843	861	880	899	918	30	19
997	30	918	937	955	974	993	*012	*030	*049	*068	*087	*105	20	19
997	40	8.05 105	124	143	161	180	199	218	236	255	274	292	10	19
997	50	292	311	329	348	367	385	404	422	441	460	478	0 **21**	19
997	**39** 0	478	497	515	534	552	571	589	608	626	645	663	50	18
997	10	663	682	700	719	737	756	774	792	811	829	848	40	18
997	20	848	866	885	903	921	940	958	976	995	*018	*031	30	18
997	30	8.06 031	050	068	086	105	123	141	159	178	196	214	20	18
997	40	214	232	251	269	287	305	324	342	360	378	396	10	18
997	50	396	414	433	451	469	487	505	523	541	560	578	0 **20**	18
9.99		10"	9"	8"	7"	6"	5"	4"	3"	2"	1"	0"	'' '	d.

L. Sin.   **89°**   L. Cos.

L. Tang. 0°

' ''	0''	1''	2''	3''	4''	5''	6''	7''	8''	9''	10''		d.	P. P.
30 0	7.94 086	110	134	158	182	206	230	254	278	302	326	50	24	
10	326	350	874	898	422	446	470	494	518	542	566	40	24	
20	566	590	618	637	661	685	709	782	756	780	804	80	24	**25**
80	804	827	851	875	899	922	946	970	993	*017	*040	20	24	1 2,5
40	7.95 040	064	088	111	185	158	182	205	229	252	276	10	24	2 5,0
50	276	299	823	346	370	893	416	440	463	487	510	0 29	23	3 7,5
31 0	510	583	557	580	608	627	650	678	696	720	748	50	28	4 10,0
10	743	766	789	812	886	859	882	905	928	951	974	40	28	5 12,5
20	974	998	*021	*044	*067	*090	*118	*186	*159	*182	*205	80	28	6 15,0
80	7.96 205	228	251	274	297	820	343	365	888	411	434	20	28	7 17,5
40	434	457	480	508	525	548	571	594	617	639	662	10	28	8 20,0
50	662	685	708	780	758	776	798	821	844	866	889	0 28	28	9 22,5
32 0	889	911	984	957	979	*002	*024	*047	*069	*092	*114	50	22	**24** **23**
10	7.97 114	187	159	182	204	227	249	272	294	817	389	50	22	1 2,4 2,3
20	889	361	884	406	428	451	473	495	518	540	562	80	22	2 4,8 4,6
80	562	585	607	629	651	673	696	718	740	762	784	20	22	3 7,2 6,9
40	784	807	829	851	878	895	917	989	961	988	*005	10	22	4 9,6 9,2
50	7.98 005	027	050	072	094	116	188	159	181	203	225	0 27	22	5 12,0 11,5
33 0	225	247	269	291	818	885	357	379	400	422	444	50	22	6 14,4 18,8
10	444	466	488	510	531	558	575	597	618	640	662	40	22	7 16,8 16,1
20	662	684	705	727	749	770	792	814	885	857	878	80	22	8 19,2 18,4
80	878	900	922	948	965	986	*008	*029	*051	*073	*094	20	22	9 21,6 20,7
40	7.99 094	116	187	158	180	201	228	244	266	287	308	10	21	**22**
50	808	830	851	873	894	415	437	458	479	501	522	0 26	21	1 2,2
34 0	522	543	564	586	607	628	649	671	692	713	784	50	21	2 4,4
10	784	755	777	798	819	840	861	882	908	925	946	40	21	3 6,6
20	946	967	988	*009	*080	*051	*072	*098	*114	*185	*156	80	21	4 8,8
80	8.00 156	177	198	219	240	261	282	808	824	844	865	20	21	5 11,0
40	865	886	407	428	449	470	490	511	582	558	574	10	21	6 18,2
50	574	594	615	636	657	677	698	719	740	760	781	0 25	21	7 15,4
35 0	781	802	822	843	964	884	905	925	946	967	987	50	21	8 17,6
10	987	*008	*028	*049	*070	*090	*111	*131	*152	*172	*193	40	21	9 19,8
20	8.01 198	218	234	254	274	295	815	886	856	377	897	80	20	**21**
80	897	417	488	458	478	499	519	589	560	580	600	20	20	1 2,1
40	600	621	641	661	682	702	722	742	762	788	808	10	20	2 4,2
50	808	828	848	868	884	904	924	944	964	984	*004	0 24	20	3 6,3
36 0	8.02 004	025	045	065	085	105	125	145	165	185	205	50	20	4 8,4
10	205	225	245	265	285	805	825	845	865	885	405	40	20	5 10,5
20	405	425	445	464	484	504	524	544	564	584	604	80	20	6 12,6
80	604	628	648	668	688	708	722	742	762	782	801	20	20	7 14,7
40	801	821	841	861	880	900	920	989	959	979	998	10	20	8 16,8
50	998	*018	*088	*057	*077	*097	*116	*186	*155	*175	*194	0 28	20	9 18,9
37 0	8.03 194	214	234	258	273	292	812	881	851	370	390	50	20	**20** **19**
10	390	409	429	448	468	487	506	526	545	565	584	40	19	1 2,0 1,9
20	584	608	628	642	661	681	700	720	789	758	777	80	19	2 4,0 8,8
80	777	797	816	885	855	874	898	912	982	951	970	20	19	3 6,0 5,7
40	970	989	*008	*028	*047	*066	*085	*104	*124	*148	*162	10	19	4 8,0 7,6
50	8.04 162	181	200	219	238	257	276	296	815	884	858	0 22	19	5 10,0 9,5
38 0	858	872	891	410	429	448	467	486	505	524	543	50	19	6 12,0 11,4
10	548	562	581	600	619	688	656	675	694	718	782	40	19	7 14,0 18,8
20	782	751	770	789	808	826	845	864	888	902	921	80	19	8 16,0 15,2
80	921	989	958	977	996	*014	*088	*052	*071	*089	*108	20	19	9 18,0 17,1
40	8.05 108	127	146	164	188	202	220	289	258	276	295	10	19	**18**
50	295	814	882	851	869	888	407	425	444	462	481	0 21	19	1 1,8
39 0	481	499	518	537	555	574	592	611	629	648	666	50	18	2 8,6
10	666	685	708	722	740	758	777	795	814	882	851	40	18	3 5,4
20	851	869	887	906	924	948	961	979	998	*016	*084	80	18	4 7,2
80	8.06 034	058	071	089	107	126	144	162	181	199	217	20	18	5 9,0
40	217	235	254	272	290	808	826	845	868	881	399	10	18	6 10,8
50	399	417	436	454	472	490	508	526	544	562	581	0 20	18	7 12,6
														8 14,4
	10''	9''	8''	7''	6''	5''	4''	8''	2''	1''	0''	'' '	d.	9 16,2

89° L. Cotg.

L. Cos. L. Sin. **0°**

9.99	′	″	0″	1″	2″	3″	4″	5″	6″	7″	8″	9″	10″		d.
997	**40**	0	8.06 578	596	614	632	650	668	686	704	722	740	758	50	18
997		10	758	776	794	812	830	848	866	884	902	920	938	40	18
997		20	938	956	974	992	∗010	∗028	∗046	∗063	∗081	∗099	∗117	30	18
997		30	8.07 117	135	153	171	189	206	224	242	260	278	295	20	18
997		40	295	313	331	349	367	384	402	420	438	455	473	10	18
997		50	473	491	509	526	544	562	579	597	615	632	650	0 19	18
997	**41**	0	650	668	685	703	721	738	756	773	791	809	826	50	18
997		10	826	844	861	879	896	914	932	949	967	984	∗002	40	18
997		20	8.08 002	019	037	054	072	089	107	124	141	159	176	30	17
997		30	176	194	211	229	246	263	281	298	316	333	350	20	17
997		40	350	368	385	403	420	437	455	472	489	506	524	10	17
997		50	524	541	558	576	593	610	627	645	662	679	696	0 18	17
997	**42**	0	696	714	731	748	765	783	800	817	834	851	868	50	17
997		10	868	886	903	920	937	954	971	988	∗006	∗023	∗040	40	17
997		20	8.09 040	057	074	091	108	125	142	159	176	193	210	30	17
997		30	210	227	244	261	278	295	312	329	346	363	380	20	17
997		40	380	397	414	431	448	465	482	499	516	533	550	10	17
997		50	550	567	583	600	617	634	651	668	685	701	718	0 17	17
997	**43**	0	718	735	752	769	786	802	819	836	853	870	886	50	17
997		10	886	903	920	937	953	970	987	∗004	∗020	∗037	∗054	40	17
997		20	8.10 054	070	087	104	120	137	154	170	187	204	220	30	17
997		30	220	237	254	270	287	303	320	337	353	370	386	20	17
996		40	386	403	420	436	453	469	486	502	519	535	552	10	17
996		50	552	568	585	601	618	634	651	667	684	700	717	0 16	16
996	**44**	0	717	733	750	766	782	799	815	832	848	864	881	50	16
996		10	881	897	914	930	946	963	979	995	∗012	∗028	∗044	40	16
996		20	8.11 044	061	077	093	110	126	142	159	175	191	207	30	16
996		30	207	224	240	256	272	289	305	321	337	354	370	20	16
996		40	370	386	402	418	435	451	467	483	499	515	531	10	16
996		50	531	548	564	580	596	612	628	644	660	677	693	0 15	16
996	**45**	0	693	709	725	741	757	773	789	805	821	837	853	50	16
996		10	853	869	885	901	917	933	949	965	981	997	∗013	40	16
996		20	8.12 013	029	045	061	077	093	109	125	141	157	172	30	16
996		30	172	188	204	220	236	252	268	284	300	315	331	20	16
996		40	331	347	363	379	395	410	426	442	458	474	489	10	16
996		50	489	505	521	537	553	568	584	600	616	631	647	0 14	16
996	**46**	0	647	663	679	694	710	726	741	757	773	788	804	50	16
996		10	804	820	836	851	867	882	898	914	929	945	961	40	16
996		20	961	976	992	∗007	∗023	∗039	∗054	∗070	∗085	∗101	∗117	30	16
996		30	8.13 117	132	148	163	179	194	210	225	241	256	272	20	16
996		40	272	287	303	318	334	349	365	380	396	411	427	10	16
996		50	427	442	458	473	489	504	519	535	550	566	581	0 13	15
996	**47**	0	581	596	612	627	643	658	673	689	704	719	735	50	15
996		10	735	750	765	781	796	811	827	842	857	873	888	40	15
996		20	888	903	919	934	949	964	980	995	∗010	∗025	∗041	30	15
996		30	8.14 041	056	071	086	101	117	132	147	162	178	193	20	15
996		40	193	208	223	238	253	269	284	299	314	329	344	10	15
996		50	344	359	375	390	405	420	435	450	465	480	495	0 12	15
996	**48**	0	495	510	525	541	556	571	586	601	616	631	646	50	15
996		10	646	661	676	691	706	721	736	751	766	781	796	40	15
996		20	796	811	826	841	856	871	886	901	915	930	945	30	15
996		30	945	960	975	990	∗005	∗020	∗035	∗050	∗065	∗079	∗094	20	15
996		40	8.15 094	109	124	139	154	169	183	198	213	228	243	10	15
996		50	243	258	272	287	302	317	332	346	361	376	391	0 11	15
996	**49**	0	391	406	420	435	450	465	479	494	509	523	538	50	15
996		10	538	553	568	582	597	612	626	641	656	670	685	40	15
996		20	685	700	714	729	744	758	773	788	802	817	832	30	15
995		30	832	846	861	875	890	905	919	934	948	963	978	20	15
995		40	978	992	∗007	∗021	∗036	∗050	∗065	∗079	∗094	∗109	∗123	10	14
995		50	8.16 123	138	152	167	181	196	210	225	239	254	268	0 10	14
9.99			10″	9″	8″	7″	6″	5″	4″	3″	2″	1″	0″	″ ′	d.

L. Sin. **89°** L. Cos.

L. Tang.   0°

' ''	0'' 1'' 2'' 3'' 4''	5''	6'' 7'' 8'' 9'' 10''		d.
40 0	8.06 581 599 617 635 653	671	689 707 725 743 761	50	18
10	761 779 797 815 833	851	869 887 905 923 941	40	18
20	941 959 977 995 *013	*031	*049 *066 *084 *102 *120	30	18
30	8.07 120 138 156 174 192	209	227 245 263 281 298	20	18
40	298 316 334 352 370	387	405 423 441 458 476	10	18
50	476 494 512 529 547	565	582 600 618 635 653	0 19	18
41 0	653 671 688 706 724	741	759 776 794 812 829	50	18
10	829 847 864 882 900	917	935 952 970 987 *005	40	18
20	8.08 005 022 040 057 075	092	110 127 145 162 180	30	18
30	180 197 214 232 249	267	284 301 319 336 354	20	17
40	354 371 388 406 423	440	458 475 492 510 527	10	17
50	527 544 562 579 596	613	631 648 665 682 700	0 18	17
42 0	700 717 734 751 769	786	803 820 837 855 872	50	17
10	872 889 906 923 940	957	975 992 *009 *026 *043	40	17
20	8.09 043 060 077 094 111	128	146 163 180 197 214	30	17
30	214 231 248 265 282	299	316 333 350 367 384	20	17
40	384 401 418 435 452	468	485 502 519 536 553	10	17
50	553 570 587 604 621	637	654 671 688 705 722	0 17	17
43 0	722 739 755 772 789	806	823 839 856 873 890	50	17
10	890 907 923 940 957	974	990 *007 *024 *040 *057	40	17
20	8.10 057 074 091 107 124	141	157 174 191 207 224	30	17
30	224 240 257 274 290	307	324 340 357 373 390	20	17
40	390 407 423 440 456	473	489 506 522 539 555	10	16
50	555 572 588 605 621	638	654 671 687 704 720	0 16	16
44 0	720 737 753 770 786	802	819 835 852 868 884	50	16
10	884 901 917 934 950	966	983 999 *015 *032 *048	40	16
20	8.11 048 064 081 097 113	130	146 162 178 195 211	30	16
30	211 227 244 260 276	292	309 325 341 357 373	20	16
40	373 390 406 422 438	454	471 487 503 519 535	10	16
50	535 551 567 584 600	616	632 648 664 680 696	0 15	16
45 0	696 712 729 745 761	777	793 809 825 841 857	50	16
10	857 873 889 905 921	937	953 969 985 *001 *017	40	16
20	8.12 017 033 049 065 081	097	113 129 144 160 176	30	16
30	176 192 208 224 240	256	272 288 303 319 335	20	16
40	335 351 367 383 398	414	430 446 462 478 493	10	16
50	493 509 525 541 556	572	588 604 620 635 651	0 14	16
46 0	651 667 682 698 714	730	745 761 777 792 808	50	16
10	808 824 839 855 871	886	902 918 933 949 965	40	16
20	965 980 996 *011 *027	*043	*058 *074 *089 *105 *121	30	16
30	8.13 121 136 152 167 183	198	214 229 245 260 276	20	16
40	276 291 307 322 338	353	369 384 400 415 431	10	16
50	431 446 462 477 493	508	523 539 554 570 585	0 13	15
47 0	585 601 616 631 647	662	677 693 708 724 739	50	15
10	739 754 770 785 800	816	831 846 861 877 892	40	15
20	892 907 923 938 953	968	984 999 *014 *029 *045	30	15
30	8.14 045 060 075 090 106	121	136 151 166 182 197	20	15
40	197 212 227 242 258	273	288 303 318 333 348	10	15
50	348 364 379 394 409	424	439 454 469 484 500	0 12	15
48 0	500 515 530 545 560	575	590 605 620 635 650	50	15
10	650 665 680 695 710	725	740 755 770 785 800	40	15
20	800 815 830 845 860	875	890 905 920 935 950	30	15
30	950 965 980 994 *009	*024	*039 *054 *069 *084 *099	20	15
40	8.15 099 114 128 143 158	173	188 203 218 232 247	10	15
50	247 262 277 292 306	321	336 351 366 380 395	0 11	15
49 0	395 410 425 439 454	469	484 498 513 528 543	50	15
10	543 557 572 587 602	616	631 646 660 675 690	40	15
20	690 704 719 734 748	763	778 792 807 822 836	30	15
30	836 851 865 880 894	909	924 938 953 968 982	20	15
40	982 997 *011 *026 *040	*055	*070 *084 *099 *113 *128	10	15
50	8.16 128 142 157 171 186	200	215 229 244 258 273	0 10	14
	10'' 9'' 8'' 7'' 6''	5''	4'' 3'' 2'' 1'' 0''	'' '	d.

P. P.

18		17		16		15		14	
1	1,8	1	1,7	1	1,6	1	1,5	1	1,4
2	3,6	2	3,4	2	3,2	2	3,0	2	2,8
3	5,4	3	5,1	3	4,8	3	4,5	3	4,2
4	7,2	4	6,8	4	6,4	4	6,0	4	5,6
5	9,0	5	8,5	5	8,0	5	7,5	5	7,0
6	10,8	6	10,2	6	9,6	6	9,0	6	8,4
7	12,6	7	11,9	7	11,2	7	10,5	7	9,8
8	14,4	8	13,6	8	12,8	8	12,0	8	11,2
9	16,2	9	15,3	9	14,4	9	13,5	9	12,6

L. Cos. L. Sin. **0°**

9.99	′ ″	0″	1″	2″	3″	4″	5″	6″	7″	8″	9″	10″		d.
995	50 0	8.16 268	288	297	311	326	340	355	369	384	398	413	50	14
995	10	413	427	441	456	470	485	499	513	528	542	557	40	14
995	20	557	571	585	600	614	628	643	657	672	686	700	30	14
995	30	700	715	729	743	757	772	786	800	815	829	843	20	14
995	40	843	858	872	886	900	915	929	943	957	972	986	10	14
995	50	986	*000	*014	*029	*043	*057	*071	*085	*100	*114	*128	0 9	14
995	51 0	8.17 128	142	156	171	185	199	213	227	241	256	270	50	14
995	10	270	284	298	312	326	340	355	369	383	397	411	40	14
995	20	411	425	439	453	467	481	495	510	524	538	552	30	14
995	30	552	566	580	594	608	622	636	650	664	678	692	20	14
995	40	692	706	720	734	748	762	776	790	804	818	832	10	14
995	50	832	846	860	874	888	902	916	930	943	957	971	0 8	14
995	52 0	971	985	999	*013	*027	*041	*055	*069	*082	*096	*110	50	14
995	10	8.18 110	124	138	152	166	180	193	207	221	235	249	40	14
995	20	249	263	276	290	304	318	332	345	359	373	387	30	14
995	30	387	401	414	428	442	456	469	483	497	511	524	20	14
995	40	524	538	552	566	579	593	607	621	634	648	662	10	14
995	50	662	675	689	703	716	730	744	757	771	785	798	0 7	14
995	53 0	798	812	826	839	853	867	880	894	908	921	935	50	14
995	10	935	948	962	976	989	*003	*016	*030	*044	*057	*071	40	14
995	20	8.19 071	084	098	111	125	139	152	166	179	193	206	30	14
995	30	206	220	233	247	260	274	287	301	314	328	341	20	14
995	40	341	355	368	382	395	409	422	436	449	463	476	10	14
995	50	476	489	503	516	530	543	557	570	583	597	610	0 6	13
995	54 0	610	624	637	650	664	677	691	704	717	731	744	50	13
995	10	744	757	771	784	797	811	824	837	851	864	877	40	13
995	20	877	891	904	917	931	944	957	971	984	997	*010	30	13
995	30	8.20 010	024	037	050	064	077	090	103	117	130	143	20	13
995	40	143	156	170	183	196	209	222	236	249	262	275	10	13
994	50	275	288	302	315	328	341	354	368	381	394	407	0 5	13
994	55 0	407	420	433	446	460	473	486	499	512	525	538	50	13
994	10	538	552	565	578	591	604	617	630	643	656	669	40	13
994	20	669	682	696	709	722	735	748	761	774	787	800	30	13
994	30	800	813	826	839	852	865	878	891	904	917	930	20	13
994	40	930	943	956	969	982	995	*008	*021	*034	*047	*060	10	13
994	50	8.21 060	073	086	099	112	125	138	151	164	177	189	0 4	13
994	56 0	189	202	215	228	241	254	267	280	293	306	319	50	13
994	10	319	331	344	357	370	383	396	409	422	434	447	40	13
994	20	447	460	473	486	499	511	524	537	550	563	576	30	13
994	30	576	588	601	614	627	640	652	665	678	691	703	20	13
994	40	703	716	729	742	754	767	780	793	805	818	831	10	13
994	50	831	844	856	869	882	895	907	920	933	945	958	0 3	13
994	57 0	958	971	983	996	*009	*022	*034	*047	*060	*072	*085	50	13
994	10	8.22 085	098	110	123	136	148	161	173	186	199	211	40	13
994	20	211	224	237	249	262	274	287	300	312	325	337	30	13
994	30	337	350	363	375	388	400	413	425	438	451	463	20	13
994	40	463	476	488	501	513	526	538	551	563	576	588	10	12
994	50	588	601	613	626	638	651	663	676	688	701	713	0 2	12
994	58 0	713	726	738	751	763	776	788	801	813	826	838	50	12
994	10	838	850	863	875	888	900	913	925	937	950	962	40	12
994	20	962	975	987	999	*012	*024	*037	*049	*061	*074	*086	30	12
994	30	8.23 086	098	111	123	136	148	160	173	185	197	210	20	12
994	40	210	222	234	247	259	271	284	296	308	321	333	10	12
994	50	333	345	357	370	382	394	407	419	431	443	456	0 1	12
994	59 0	456	468	480	492	505	517	529	541	554	566	578	50	12
994	10	578	590	603	615	627	639	652	664	676	688	700	40	12
994	20	700	713	725	737	749	761	773	786	798	810	822	30	12
993	30	822	834	846	859	871	883	895	907	919	931	944	20	12
993	40	944	956	968	980	992	*004	*016	*028	*041	*053	*065	10	12
993	50	8.24 065	077	089	101	113	125	137	149	161	173	186	0 0	12
9.99		10″	9″	8″	7″	6″	5″	4″	3″	2″	1″	0″	″ ′	d.

L. Sin. **89°** L. Cos.

L. Tang.   0°

' ''	0'' 1'' 2'' 3'' 4''	5''	6'' 7'' 8'' 9'' 10''		d.	P. P.
50 0	8.16 278 267 302 316 331	345	359 374 388 403 417	50	14	
10	417 432 446 460 475	489	504 518 533 547 561	40	14	
20	561 576 590 604 619	633	647 662 676 691 705	30	14	
30	705 719 734 748 762	776	791 805 819 834 848	20	14	
40	848 863 877 891 905	919	934 948 962 976 991	10	14	
50	991 ₊005 ₊019 ₊033 ₊048	₊062	₊076 ₊090 ₊104 ₊119 ₊133	0 9	14	**15**
51 0	8.17 133 147 161 175 190	204	218 232 246 260 275	50	14	1| 1,5
10	275 289 303 317 331	345	359 373 388 402 416	40	14	2| 3,0
20	416 430 444 458 472	486	500 514 528 543 557	30	14	3| 4,5
30	557 571 585 599 613	627	641 655 669 688 697	20	14	4| 6,0
40	697 711 725 739 753	767	781 795 809 823 837	10	14	5| 7,5
50	837 851 865 879 893	907	921 934 948 962 976	0 8	14	6| 9,0 7| 10,5 8| 12,0 9| 13,5
52 0	976 990 ₊004 ₊018 ₊032	₊046	₊060 ₊074 ₊087 ₊101 ₊115	50	14	
10	8.18 115 129 143 157 171	185	198 212 226 240 254	40	14	
20	254 268 281 295 309	323	337 351 364 378 392	30	14	
30	392 406 419 433 447	461	475 488 502 516 530	20	14	
40	530 543 557 571 585	598	612 626 639 653 667	10	14	
50	667 681 694 708 722	735	749 763 776 790 804	0 7	14	
53 0	804 817 831 845 858	872	886 899 913 926 940	50	14	**14**
10	940 954 967 981 994	₊008	₊022 ₊035 ₊049 ₊062 ₊076	40	14	
20	8.19 076 090 103 117 130	144	157 171 184 198 211	30	14	1| 1,4
30	211 225 239 252 266	279	293 306 320 333 347	20	14	2| 2,8
40	347 360 374 387 401	414	427 441 454 468 481	10	13	3| 4,2
50	481 495 508 522 535	548	562 575 589 602 616	0 6	13	4| 5,6 5| 7,0
54 0	616 629 642 656 669	683	696 709 723 736 749	50	13	6| 8,4
10	749 763 776 789 803	816	830 843 856 870 883	40	13	7| 9,8
20	883 896 910 923 936	949	963 976 989 ₊003 ₊016	30	13	8| 11,2
30	8.20 016 029 042 056 069	082	096 109 122 135 149	20	13	9| 12,6
40	149 162 175 188 201	215	228 241 254 268 281	10	13	
50	281 294 307 320 334	347	360 373 386 399 413	0 5	13	
55 0	413 426 439 452 465	478	491 505 518 531 544	50	13	
10	544 557 570 583 596	610	623 636 649 662 675	40	13	
20	675 688 701 714 727	740	753 767 780 793 806	30	13	
30	806 819 832 845 858	871	884 897 910 923 936	20	13	**13**
40	936 949 962 975 988	₊001	₊014 ₊027 ₊040 ₊053 ₊066	10	13	
50	8.21 066 079 092 105 118	131	144 156 169 182 195	0 4	13	1| 1,3
56 0	195 208 221 234 247	260	273 286 299 311 324	50	13	2| 2,6
10	324 337 350 363 376	389	402 414 427 440 453	40	13	3| 3,9
20	453 466 479 492 504	517	530 543 556 569 581	30	13	4| 5,2
30	581 594 607 620 633	645	658 671 684 697 709	20	13	5| 6,5 6| 7,8
40	709 722 735 748 760	773	786 799 811 824 837	10	13	7| 9,1
50	837 850 862 875 888	901	913 926 939 951 964	0 3	13	8| 10,4 9| 11,7
57 0	964 977 989 ₊002 ₊015	₊028	₊040 ₊053 ₊066 ₊078 ₊091	50	13	
10	8.22 091 104 116 129 142	154	167 179 192 205 217	40	13	
20	217 230 243 255 268	280	293 306 318 331 343	30	13	
30	343 356 369 381 394	406	419 431 444 457 469	20	13	
40	469 482 494 507 519	532	544 557 569 582 595	10	13	
50	595 607 620 632 645	657	670 682 695 707 720	0 2	12	**12**
58 0	720 732 744 757 769	782	794 807 819 832 844	50	12	1| 1,2
10	844 857 869 881 894	906	919 931 944 956 968	40	12	2| 2,4
20	968 981 993 ₊006 ₊018	₊030	₊043 ₊055 ₊068 ₊080 ₊092	30	12	3| 3,6
30	8.23 092 105 117 130 142	154	167 179 191 204 216	20	12	4| 4,8
40	216 228 241 253 265	278	290 302 315 327 339	10	12	5| 6,0 6| 7,2
50	339 352 364 376 388	401	413 425 438 450 462	0 1	12	7| 8,4
59 0	462 474 487 499 511	523	536 548 560 572 585	50	12	8| 9,6 9| 10,8
10	585 597 609 621 634	646	658 670 682 695 707	40	12	
20	707 719 731 743 756	768	780 792 804 816 829	30	12	
30	829 841 853 865 877	889	902 914 926 938 950	20	12	
40	950 962 974 987 999	₊011	₊023 ₊035 ₊047 ₊059 ₊071	10	12	
50	8.24 071 083 096 108 120	132	144 156 168 180 192	0 0	12	
	10'' 9'' 8'' 7'' 6''	5''	4'' 3'' 2'' 1'' 0''	'' '	d.	P. P.

89° L. Cotg.

9.99	′	0″	10″	20″	30″	40″	50″	60″		d.	P. P.
998	0	8.24 186	306	426	546	665	785	903	59	120	
998	1	903	*022	*140	*258	*375	*493	*609	58	118	120 119 118
998	2	8.25 609	726	842	958	*074	*189	*304	57	116	1| 12.0 11.9 11.8
998	3	8.26 804	419	588	648	761	875	988	56	114	2| 24.0 23.8 23.6
992	4	988	*101	*214	*326	*438	*550	*661	55	112	3| 36.0 35.7 35.4
											4| 48.0 47.6 47.2
992	5	8.27 661	773	883	994	*104	*215	*324	54	110	5| 60.0 59.5 59.0
992	6	8.28 324	434	543	652	761	869	977	53	109	6| 72.0 71.4 70.8
992	7	977	*085	*193	*300	*407	*514	*621	52	107	7| 84.0 83.3 82.6
992	8	8.29 621	727	833	939	*044	*150	*255	51	106	8| 96.0 95.2 94.4
991	9	8.30 255	359	464	568	672	776	879	50	104	9| 108.0 107.1 106.2
991	10	879	983	*086	*188	*291	*393	*495	49	103	117 116 115
991	11	8.31 495	597	699	800	901	*002	*103	48	101	1| 11.7 11.6 11.5
990	12	8.32 103	206	303	408	503	602	702	47	100	2| 23.4 23.2 23.0
990	13	702	801	899	993	*096	*195	*292	46	98	3| 35.1 34.8 34.5
990	14	8.33 292	390	488	585	682	779	875	45	97	4| 46.8 46.4 46.0
											5| 58.5 58.0 57.5
990	15	875	972	*068	*164	*260	*355	*450	44	96	6| 70.2 69.6 69.0
989	16	8.34 450	546	640	735	830	924	*018	43	95	7| 81.9 81.2 80.5
989	17	8.35 018	112	206	299	392	485	578	42	93	8| 93.6 92.8 92.0
989	18	578	671	764	856	948	*040	*131	41	92	9| 105.3 104.4 103.5
989	19	8.36 181	223	314	405	496	587	678	40	91	
											114 113 112 111
988	20	678	768	858	948	*038	*128	*217	39	90	1| 11.4 11.3 11.2 11.1
988	21	8.37 217	306	395	484	573	662	750	38	89	2| 22.8 22.6 22.4 22.2
988	22	750	838	926	*014	*101	*189	*276	37	88	3| 34.2 33.9 33.6 33.3
987	23	8.38 276	363	450	537	624	710	796	36	87	4| 45.6 45.2 44.8 44.4
987	24	796	882	968	*054	*139	*225	*310	35	86	5| 57.0 56.5 56.0 55.5
											6| 68.4 67.8 67.2 66.6
987	25	8.39 310	395	480	565	649	734	818	34	85	7| 79.8 79.1 78.4 77.7
986	26	818	902	986	*070	*153	*237	*320	33	84	8| 91.2 90.4 89.6 88.8
986	27	8.40 320	403	486	569	651	734	816	32	83	9| 102.6 101.7 100.8 99.9
986	28	816	898	980	*062	*144	*225	*307	31	82	
985	29	8.41 307	388	469	550	631	711	792	30	81	110 109 108 107
											1| 11.0 10.9 10.8 10.7
985	30	792	872	952	*032	*112	*192	*272	29	80	2| 22.0 21.8 21.6 21.4
985	31	8.42 272	351	430	510	589	667	746	28	79	3| 33.0 32.7 32.4 32.1
984	32	746	825	903	982	*060	*138	*216	27	78	4| 44.0 43.6 43.2 42.8
984	33	8.43 216	293	371	448	526	603	680	26	77	5| 55.0 54.5 54.0 53.5
984	34	680	757	834	910	987	*063	*139	25	76	6| 66.0 65.4 64.8 64.2
											7| 77.0 76.3 75.6 74.9
983	35	8.44 189	216	292	367	443	519	594	24	76	8| 88.0 87.2 86.4 85.6
983	36	594	669	745	820	895	969	*044	23	75	9| 99.0 98.1 97.2 96.3
983	37	8.45 044	119	193	267	341	415	489	22	74	
982	38	489	563	637	710	784	857	930	21	74	106 105 104 103
982	39	930	*003	*076	*149	*222	*294	*366	20	73	1| 10.6 10.5 10.4 10.3
											2| 21.2 21.0 20.8 20.6
982	40	8.46 366	439	511	583	655	727	799	19	72	3| 31.8 31.5 31.2 30.9
981	41	799	870	942	*013	*084	*155	*226	18	71	4| 42.4 42.0 41.6 41.2
981	42	8.47 226	297	368	489	509	580	650	17	71	5| 53.0 52.5 52.0 51.5
981	43	650	720	790	860	930	*000	*069	16	70	6| 63.6 63.0 62.4 61.8
980	44	8.48 069	139	208	278	347	416	485	15	69	7| 74.2 73.5 72.8 72.1
											8| 84.8 84.0 83.2 82.4
980	45	485	554	622	691	760	828	896	14	68	9| 95.4 94.5 93.6 92.7
979	46	896	965	*088	*101	*169	*236	*304	13	68	
979	47	8.49 304	372	439	506	574	641	708	12	67	102 101 100 99
979	48	708	775	842	908	975	*042	*108	11	67	1| 10.2 10.1 10.0 9.9
978	49	8.50 103	174	241	307	373	489	504	10	66	2| 20.4 20.2 20.0 19.8
											3| 30.6 30.3 30.0 29.7
978	50	504	570	636	701	767	832	897	9	66	4| 40.8 40.4 40.0 39.6
977	51	897	963	*028	*092	*157	*222	*287	8	65	5| 51.0 50.5 50.0 49.5
977	52	8.51 287	351	416	480	544	609	673	7	64	6| 61.2 60.6 60.0 59.4
977	53	678	737	801	864	928	992	*055	6	64	7| 71.4 70.7 70.0 69.3
976	54	8.52 055	119	182	245	808	871	484	5	63	8| 81.6 80.8 80.0 79.2
											9| 91.8 90.9 90.0 89.1
976	55	484	497	560	623	685	748	810	4	63	
975	56	810	872	935	997	*059	*121	*183	3	62	98 97 96 95
975	57	8.53 183	245	306	368	429	491	552	2	62	1| 9.8 9.7 9.6 9.5
974	58	552	614	675	736	797	858	919	1	61	2| 19.6 19.4 19.2 19.0
974	59	919	979	*040	*101	*161	*222	*282	0	60	3| 29.4 29.1 28.8 28.5
											4| 39.2 38.8 38.4 38.0
											5| 49.0 48.5 48.0 47.5
											6| 58.8 58.2 57.6 57.0
											7| 68.6 67.9 67.2 66.5
											8| 78.4 77.6 76.8 76.0
											9| 88.2 87.3 86.4 85.5
9.99		60″	50″	40″	30″	20″	10″	0″	′	d.	P. P.

L. Tang.   1°

′	0″	10″	20″	30″	40″	50″	60″	·	d.
0	8.24 192	318	438	558	672	791	910	59	120
1	910	*029	*147	*265	*382	*500	*616	58	118
2	8.25 616	738	849	965	*081	*196	*312	57	116
3	8.26 812	426	541	655	769	882	996	56	114
4	996	*109	*221	*334	*446	*558	*669	55	112
5	8.27 669	780	891	*002	*112	*228	*332	54	110
6	8.28 332	442	551	660	769	877	986	53	109
7	986	*094	*201	*309	*416	*523	*629	52	107
8	8.29 629	736	842	947	*053	*158	*263	51	106
9	8.30 263	368	478	577	681	785	888	50	104
10	888	992	*095	*198	*300	*403	*505	49	103
11	8.31 505	606	708	809	911	*012	*112	48	101
12	8.32 112	213	313	413	513	612	711	47	100
13	711	810	909	*008	*106	*205	*302	46	98
14	8.33 302	400	498	595	692	789	886	45	97
15	886	982	*078	*174	*270	*366	*461	44	96
16	8.34 461	556	651	746	840	985	*029	43	95
17	8.35 029	123	217	310	403	497	590	42	94
18	590	682	775	867	959	*051	*143	41	92
19	8.36 143	235	326	417	508	599	689	40	91
20	689	780	870	960	*050	*140	*229	39	90
21	8.37 229	318	408	497	585	674	762	38	89
22	762	850	938	*026	*114	*202	*289	37	88
23	8.38 289	376	463	550	636	723	809	36	87
24	809	895	981	*067	*153	*238	*323	35	86
25	8.39 323	408	493	578	663	747	832	34	85
26	832	916	*000	*083	*167	*250	*334	33	84
27	8.40 384	417	500	583	665	748	830	32	82
28	830	913	995	*077	*158	*240	*321	31	82
29	8.41 321	403	484	565	646	726	807	30	81
30	807	887	967	*048	*127	*207	*287	29	80
31	8.42 287	366	446	525	604	683	762	28	79
32	762	840	919	997	*075	*154	*232	27	78
33	8.43 232	309	387	464	542	619	696	26	77
34	696	773	850	927	*008	*080	*156	25	77
35	8.44 156	232	308	384	460	536	611	24	76
36	611	686	762	837	912	987	*061	23	75
37	8.45 061	186	210	285	359	433	507	22	74
38	507	581	655	728	802	875	948	21	74
39	948	*021	*094	*167	*240	*312	*385	20	73
40	8.46 385	457	529	602	674	745	817	19	72
41	817	889	960	*082	*103	*174	*245	18	71
42	8.47 245	316	387	458	528	599	669	17	71
43	669	740	810	880	950	*020	*089	16	70
44	8.48 089	159	228	298	367	436	505	15	69
45	505	574	643	711	780	849	917	14	69
46	917	985	*053	*121	*189	*257	*325	13	68
47	8.49 325	393	460	528	595	662	729	12	67
48	729	796	863	930	997	*068	*180	11	67
49	8.50 180	196	263	329	395	461	527	10	66
50	527	593	658	724	789	855	920	9	66
51	920	985	*050	*115	*180	*245	*310	8	65
52	8.51 310	374	439	503	568	632	696	7	64
53	696	760	824	888	952	*015	*079	6	64
54	8.52 079	143	206	269	332	396	459	5	63
55	459	522	584	647	710	772	835	4	63
56	835	897	960	*022	*064	*146	*208	3	62
57	8.53 208	270	332	393	455	516	578	2	62
58	578	639	700	762	823	884	945	1	61
59	945	*005	*066	*127	*187	*248	*308	0	60
	60″	50″	40″	30″	20″	10″	0″	′	d.

P. P.

	94	93	92	91	90
1	9,4	9,3	9,2	9,1	9,0
2	18,8	18,6	18,4	18,2	18,0
3	28,2	27,9	27,6	27,3	27,0
4	37,6	37,2	36,8	36,4	36,0
5	47,0	46,5	46,0	45,5	45,0
6	56,4	55,8	55,2	54,6	54,0
7	65,8	65,1	64,4	63,7	63,0
8	75,2	74,4	73,6	72,8	72,0
9	84,6	83,7	82,8	81,9	81,0

	89	88	87	86	85
1	8,9	8,8	8,7	8,6	8,5
2	17,8	17,6	17,4	17,2	17,0
3	26,7	26,4	26,1	25,8	25,5
4	35,6	35,2	34,8	34,4	34,0
5	44,5	44,0	43,5	43,0	42,5
6	53,4	52,8	52,2	51,6	51,0
7	62,3	61,6	60,9	60,2	59,5
8	71,2	70,4	69,6	68,8	68,0
9	80,1	79,2	78,3	77,4	76,5

	84	83	82	81	80
1	8,4	8,3	8,2	8,1	8,0
2	16,8	16,6	16,4	16,2	16,0
3	25,2	24,9	24,6	24,3	24,0
4	33,6	33,2	32,8	32,4	32,0
5	42,0	41,5	41,0	40,5	40,0
6	50,4	49,8	49,2	48,6	48,0
7	58,8	58,1	57,4	56,7	56,0
8	67,2	66,4	65,6	64,8	64,0
9	75,6	74,7	73,8	72,9	72,0

	79	78	77	76	75
1	7,9	7,8	7,7	7,6	7,5
2	15,8	15,6	15,4	15,2	15,0
3	23,7	23,4	23,1	22,8	22,5
4	31,6	31,2	30,8	30,4	30,0
5	39,5	39,0	38,5	38,0	37,5
6	47,4	46,8	46,2	45,6	45,0
7	55,3	54,6	53,9	53,2	52,5
8	63,2	62,4	61,6	60,8	60,0
9	71,1	70,2	69,3	68,4	67,5

	74	73	72	71	70
1	7,4	7,3	7,2	7,1	7,0
2	14,8	14,6	14,4	14,2	14,0
3	22,2	21,9	21,6	21,3	21,0
4	29,6	29,2	28,8	28,4	28,0
5	37,0	36,5	36,0	35,5	35,0
6	44,4	43,8	43,2	42,6	42,0
7	51,8	51,1	50,4	49,7	49,0
8	59,2	58,4	57,6	56,8	56,0
9	66,6	65,7	64,8	63,9	63,0

	69	68	67	66	65
1	6,9	6,8	6,7	6,6	6,5
2	13,8	13,6	13,4	13,2	13,0
3	20,7	20,4	20,1	19,8	19,5
4	27,6	27,2	26,8	26,4	26,0
5	34,5	34,0	33,5	33,0	32,5
6	41,4	40,8	40,2	39,6	39,0
7	48,3	47,6	46,9	46,2	45,5
8	55,2	54,4	53,6	52,8	52,0
9	62,1	61,2	60,3	59,4	58,5

	64	63	62	61	60
1	6,4	6,3	6,2	6,1	6,0
2	12,8	12,6	12,4	12,2	12,0
3	19,2	18,9	18,6	18,3	18,0
4	25,6	25,2	24,8	24,4	24,0
5	32,0	31,5	31,0	30,5	30,0
6	38,4	37,8	37,2	36,6	36,0
7	44,8	44,1	43,4	42,7	42,0
8	51,2	50,4	49,6	48,8	48,0
9	57,6	56,7	55,8	54,9	54,0

9.99	′	0″	10″	20″	30″	40″	50″	60″			d.
974	0	8.54 282	842	402	462	522	582	642	59	978	60
973	1	642	702	762	821	881	940	999	58	978	60
973	2	999	*059	*118	*177	*236	*295	*354	57	972	59
972	3	8.55 354	413	471	530	589	647	705	56	972	58
972	4	705	764	822	880	938	996	*054	55	971	58
971	5	8.56 054	112	170	227	285	342	400	54	971	58
971	6	400	457	515	572	629	686	743	53	970	57
970	7	743	800	857	914	970	*027	*084	52	970	57
970	8	8.57 084	140	196	253	309	365	421	51	969	56
969	9	421	477	533	589	645	701	757	50	969	56
969	10	757	812	868	923	979	*084	*089	49	968	55
968	11	8.58 069	144	200	255	310	364	419	48	968	55
968	12	419	474	529	583	638	693	747	47	967	55
967	13	747	801	856	910	964	*018	*072	46	967	54
967	14	8.59 072	126	180	234	288	341	395	45	967	54
967	15	395	448	502	555	609	662	715	44	966	53
966	16	715	768	821	874	927	980	*033	43	966	53
966	17	8.60 033	086	189	191	244	296	349	42	965	53
965	18	349	401	454	506	558	610	662	41	964	52
964	19	662	714	766	818	870	922	973	40	964	52
964	20	973	*025	*077	*128	*180	*231	*282	39	963	52
963	21	8.61 282	334	385	436	487	538	589	38	963	51
963	22	589	640	691	742	792	843	894	37	962	51
962	23	894	944	995	*045	*096	*146	*196	36	962	50
962	24	8.62 196	246	297	347	397	447	497	35	961	50
961	25	497	546	596	646	696	745	795	34	961	50
961	26	795	844	894	943	993	*042	*091	33	960	49
960	27	8.63 091	140	189	238	288	336	385	32	960	49
960	28	385	434	483	532	580	629	678	31	959	49
959	29	678	726	775	823	871	920	968	30	959	48
959	30	968	*016	*064	*112	*160	*208	*256	29	958	48
958	31	8.64 256	304	352	400	448	495	543	28	958	48
958	32	543	590	638	685	733	780	827	27	957	47
957	33	827	875	922	969	*016	*063	*110	26	956	47
956	34	8.65 110	157	204	251	298	344	391	25	956	47
956	35	391	438	484	531	577	624	670	24	955	46
955	36	670	717	763	809	855	901	947	23	955	46
955	37	947	994	*040	*085	*131	*177	*223	22	954	46
954	38	8.66 223	269	314	360	406	451	497	21	954	46
954	39	497	542	588	633	678	724	769	20	953	45
953	40	769	814	859	904	949	994	*039	19	952	45
952	41	8.67 039	084	129	174	219	263	308	18	952	45
952	42	308	353	397	442	486	531	575	17	951	44
951	43	575	619	664	708	752	796	841	16	951	44
951	44	841	885	929	973	*017	*060	*104	15	950	44
950	45	8.68 104	148	192	236	279	323	367	14	949	44
949	46	367	410	454	497	540	584	627	13	949	43
949	47	627	670	714	757	800	843	886	12	948	43
948	48	886	929	972	*015	*058	*101	*144	11	948	43
948	49	8.69 144	187	229	272	315	357	400	10	947	43
947	50	400	442	485	527	570	612	654	9	946	42
946	51	654	697	739	781	823	865	907	8	946	42
946	52	907	949	991	*033	*075	*117	*159	7	945	42
945	53	8.70 159	201	242	284	326	367	409	6	944	42
944	54	409	451	492	534	575	616	658	5	944	42
944	55	658	699	740	781	823	864	905	4	943	41
943	56	905	946	987	*028	*069	*110	*151	3	942	41
942	57	8.71 151	192	232	273	314	355	395	2	942	41
942	58	395	436	476	517	557	598	638	1	941	40
941	59	638	679	719	759	800	840	880	0	940	40
		60″	50″	40″	30″	20″	10″	0″	′	9.99	d.

P. P.

61		60		59		58		57		56	
1	6,1	1	6,0	1	5,9	1	5,8	1	5,7	1	5,6
2	12,2	2	12,0	2	11,8	2	11,6	2	11,4	2	11,2
3	18,3	3	18,0	3	17,7	3	17,4	3	17,1	3	16,8
4	24,4	4	24,0	4	23,6	4	23,2	4	22,8	4	22,4
5	30,5	5	30,0	5	29,5	5	29,0	5	28,5	5	28,0
6	36,6	6	36,0	6	35,4	6	34,8	6	34,2	6	33,6
7	42,7	7	42,0	7	41,3	7	40,6	7	39,9	7	39,2
8	48,8	8	48,0	8	47,2	8	46,4	8	45,6	8	44,8
9	54,9	9	54,0	9	53,1	9	52,2	9	51,3	9	50,4

L. Tang.   2°

′	0″	10″	20″	30″	40″	50″	60″		d.
0	8.54 308	869	429	489	549	609	669	59	60
1	669	729	789	848	908	967	*027	58	60
2	8.55 027	086	145	205	264	828	382	57	59
3	382	441	499	558	617	675	734	56	59
4	734	792	850	909	967	*025	*088	55	58
5	8.56 088	141	199	256	314	372	429	54	58
6	429	487	544	601	659	716	773	58	57
7	778	880	887	944	*000	*057	*114	52	57
8	8.57 114	170	227	288	840	896	452	51	56
9	452	508	564	620	676	732	788	50	56
10	788	848	899	955	*010	*065	*121	49	56
11	8.58 121	176	231	286	341	896	451	48	55
12	451	506	561	616	670	725	779	47	55
13	779	884	888	948	997	*051	*105	46	54
14	8.59 105	159	213	267	321	875	428	45	54
15	428	482	536	589	642	696	749	44	54
16	749	802	856	909	962	*015	*068	48	53
17	8.60 068	121	178	226	279	881	384	42	58
18	384	436	489	541	593	646	698	41	52
19	698	750	802	854	906	958	*009	40	52
20	8.61 009	061	113	164	216	267	319	89	52
21	819	870	422	478	524	575	626	88	51
22	626	677	728	779	880	881	931	87	51
23	931	982	*083	*088	*184	*184	*284	86	50
24	8.62 284	285	385	885	485	485	585	85	50
25	585	585	635	685	735	784	884	84	50
26	834	884	933	988	*032	*061	*131	88	50
27	8.63 181	180	229	278	328	877	426	82	49
28	426	475	528	572	621	670	718	81	49
29	718	767	816	864	918	961	*009	80	48
30	8.64 009	058	106	154	202	250	298	29	48
31	298	846	894	442	490	588	585	28	48
32	585	638	681	728	776	823	870	27	48
33	870	918	965	*012	*060	*107	*154	26	47
34	8.65 154	201	248	295	342	888	485	25	47
35	485	482	529	575	622	668	715	24	47
36	715	761	808	854	900	947	993	28	46
37	998	*089	*085	*131	*177	*223	*269	22	46
38	8.66 269	815	361	406	452	498	548	21	46
39	548	589	634	680	725	771	816	20	46
40	816	861	906	952	997	*042	*087	19	45
41	8.67 087	182	177	222	267	812	356	18	45
42	856	401	446	490	585	579	624	17	45
43	624	668	713	757	801	846	890	16	44
44	890	984	978	*022	*066	*110	*154	15	44
45	8.68 154	198	242	286	880	873	417	14	44
46	417	461	504	548	592	685	678	18	44
47	678	722	765	808	852	895	988	12	43
48	988	981	*024	*067	*110	*153	*196	11	43
49	8.69 196	239	282	325	868	410	453	10	43
50	458	496	588	581	628	666	708	9	42
51	708	750	798	885	877	920	962	8	42
52	962	*004	*046	*088	*180	*172	*214	7	42
53	8.70 214	256	298	389	881	423	465	6	42
54	465	506	548	589	631	673	714	5	42
55	714	755	797	838	879	921	962	4	41
56	962	*008	*044	*085	*126	*167	*208	8	41
57	8.71 208	249	290	881	872	413	453	2	41
58	458	494	585	575	616	657	697	1	41
59	697	788	778	819	859	899	940	0	40
	60″	50″	40″	30″	20″	10″	0″	′	d.

P. P.

	55	54	53
1	5,5	5,4	5,8
2	11,0	10,8	10,6
8	16,5	16,2	15,9
4	22,0	21,6	21,2
5	27,5	27,0	26,5
6	88,0	82,4	81,8
7	88,5	87,8	87,1
8	44,0	48,2	42,4
9	49,5	48,6	47,7

	52	51
1	5,2	5,1
2	10,4	10,2
8	15,6	15,8
4	20,8	20,4
5	26,0	25,5
6	81,2	80,6
7	86,4	85,7
8	41,6	40,8
9	46,8	45,9

	50	49	48
1	5,0	4,9	4,8
2	10,0	9,8	9,6
8	15,0	14,7	14,4
4	20,0	19,6	19,2
5	25,0	24,5	24,0
6	30,0	29,4	28,8
7	85,0	84,8	88,6
8	40,0	89,2	88,4
9	45,0	44,1	48,2

	47	46	45
1	4,7	4,6	4,5
2	9,4	9,2	9,0
8	14,1	13,8	18,5
4	18,8	18,4	18,0
5	28,5	28,0	22,5
6	28,2	27,6	27,0
7	82,9	82,2	81,5
8	87,6	86,8	86,0
9	42,8	41,4	40,5

	44	43
1	4,4	4,8
2	8,8	8,6
8	13,2	12,9
4	17,6	17,2
5	22,0	21,5
6	26,4	25,8
7	80,8	80,1
8	85,2	84,4
9	89,6	88,7

	42	41	40
1	4,2	4,1	4,0
2	8,4	8,2	8,0
8	12,6	12,8	12,0
4	16,8	16,4	16,0
5	21,0	20,5	20,0
6	25,2	24,6	24,0
7	29,4	28,7	28,0
8	88,6	82,8	82,0
9	87,8	86,9	86,0

P. P.

87°   L. Cotg.

9.99	′	0″	10″	20″	30″	40″	50″	60″		9.99	d.
940	0	8.71 880	920	960	*000	*040	*080	*120	59	940	40
940	1	8.72 120	160	200	240	280	320	359	58	939	40
939	2	359	399	439	478	518	558	597	57	938	40
938	3	597	637	676	716	755	794	834	56	938	40
938	4	834	873	912	951	991	*030	*069	55	937	39
987	5	8.73 069	108	147	186	225	264	303	54	936	39
986	6	303	342	380	419	458	497	535	53	936	39
986	7	535	574	613	651	690	728	767	52	935	39
985	8	767	805	844	882	920	959	997	51	934	38
984	9	997	*035	*073	*112	*150	*188	*226	50	934	38
984	10	8.74 226	264	302	340	378	416	454	49	933	38
933	11	454	491	529	567	605	642	680	48	932	38
932	12	680	718	755	793	831	868	906	47	932	38
932	13	906	943	980	*018	*055	*092	*130	46	931	37
931	14	8.75 130	167	204	241	279	316	353	45	930	37
930	15	353	390	427	464	501	538	575	44	929	37
929	16	575	612	648	685	722	759	795	43	929	37
929	17	795	832	869	905	942	979	*015	42	928	37
928	18	8.76 015	052	088	125	161	197	234	41	927	36
927	19	234	270	306	343	379	415	451	40	926	36
926	20	451	487	523	559	595	631	667	39	926	36
926	21	667	703	739	775	811	847	883	38	925	36
925	22	883	919	954	990	*026	*061	*097	37	924	36
924	23	8.77 097	133	168	204	239	275	310	36	923	36
923	24	310	346	381	416	452	487	522	35	923	35
923	25	522	558	593	628	663	698	733	34	922	35
922	26	733	768	803	838	873	908	943	33	921	35
921	27	943	978	*013	*048	*083	*118	*152	32	920	35
920	28	8.78 152	187	222	257	291	326	360	31	920	35
920	29	360	395	430	464	499	533	568	30	919	35
919	30	568	602	636	671	705	739	774	29	918	34
918	31	774	808	842	876	910	945	979	28	917	34
917	32	979	*013	*047	*081	*115	*149	*183	27	917	34
917	33	8.79 183	217	251	284	318	352	386	26	916	34
916	34	386	420	453	487	521	555	588	25	915	34
915	35	588	622	655	689	722	756	789	24	914	34
914	36	789	823	856	890	923	956	990	23	913	34
913	37	990	*023	*056	*090	*123	*156	*189	22	913	33
913	38	8.80 189	222	255	289	322	355	388	21	912	33
912	39	388	421	454	487	519	552	585	20	911	33
911	40	585	618	651	684	716	749	782	19	910	33
910	41	782	815	847	880	913	945	978	18	909	33
909	42	978	*010	*043	*075	*108	*140	*173	17	909	32
909	43	8.81 173	205	237	270	302	334	367	16	908	32
908	44	367	399	431	463	496	528	560	15	907	32
907	45	560	592	624	656	688	720	752	14	906	32
906	46	752	784	816	848	880	912	944	13	905	32
905	47	944	975	*007	*039	*071	*103	*134	12	904	32
904	48	8.82 184	166	198	229	261	292	324	11	904	32
904	49	324	356	387	419	450	482	513	10	903	32
903	50	513	544	576	607	639	670	701	9	902	31
902	51	701	732	764	795	826	857	888	8	901	31
901	52	888	920	951	982	*013	*044	*075	7	900	31
900	53	8.83 075	106	137	168	199	230	261	6	899	31
899	54	261	292	322	353	384	415	446	5	898	31
898	55	446	476	507	538	568	599	630	4	898	31
898	56	630	660	691	721	752	783	813	3	897	30
897	57	813	844	874	904	935	965	996	2	896	30
896	58	996	*026	*056	*087	*117	*147	*177	1	895	30
895	59	8.84 177	208	238	268	298	328	358	0	894	30
		60″	50″	40″	30″	20″	10″	0″	′	9.99	d.

P. P.

	40	39
1	4,0	3,9
2	8,0	7,8
3	12,0	11,7
4	16,0	15,6
5	20,0	19,5
6	24,0	23,4
7	28,0	27,3
8	32,0	31,2
9	36,0	35,1

	38	37
1	3,8	3,7
2	7,6	7,4
3	11,4	11,1
4	15,2	14,8
5	19,0	18,5
6	22,8	22,2
7	26,6	25,9
8	30,4	29,6
9	34,2	33,3

	36
1	3,6
2	7,2
3	10,8
4	14,4
5	18,0
6	21,6
7	25,2
8	28,8
9	32,4

	35	34
1	3,5	3,4
2	7,0	6,8
3	10,5	10,2
4	14,0	13,6
5	17,5	17,0
6	21,0	20,4
7	24,5	23,8
8	28,0	27,2
9	31,5	30,6

	33	32
1	3,3	3,2
2	6,6	6,4
3	9,9	9,6
4	13,2	12,8
5	16,5	16,0
6	19,8	19,2
7	23,1	22,4
8	26,4	25,6
9	29,7	28,8

	31	30
1	3,1	3,0
2	6,2	6,0
3	9,3	9,0
4	12,4	12,0
5	15,5	15,0
6	18,6	18,0
7	21,7	21,0
8	24,8	24,0
9	27,9	27,0

L. Tang.   3°

'	0"	10"	20"	30"	40"	50"	60"		d.
0	8.71 940	980	*020	*060	*100	*141	*181	59	40
1	8.72 181	221	261	301	341	380	420	58	40
2	490	460	500	540	579	619	659	57	40
3	659	698	738	777	817	856	896	56	40
4	896	935	975	*014	*053	*093	*132	55	39
5	8.73 132	171	210	249	288	327	366	54	39
6	366	405	444	488	522	561	600	53	39
7	600	638	677	716	754	793	832	52	39
8	832	870	909	947	986	*024	*063	51	38
9	8.74 063	101	139	178	216	254	292	50	38
10	292	330	369	407	445	483	521	49	38
11	521	559	597	634	672	710	748	48	38
12	748	786	823	861	899	936	974	47	38
13	974	*012	*049	*087	*124	*162	*199	46	38
14	8.75 199	236	274	311	348	385	423	45	37
15	423	460	497	534	571	608	645	44	37
16	645	682	719	756	793	830	867	43	37
17	867	904	940	977	*014	*051	*087	42	37
18	8.76 087	124	160	197	233	270	306	41	36
19	306	343	379	416	452	488	525	40	36
20	525	561	597	633	669	706	742	39	36
21	742	778	814	850	886	922	958	38	36
22	958	994	*030	*065	*101	*137	*173	37	36
23	8.77 173	208	244	280	315	351	387	36	36
24	387	422	458	493	529	564	600	35	36
25	600	635	670	706	741	776	811	34	35
26	811	847	882	917	952	987	*022	33	35
27	8.78 022	057	092	127	162	197	232	32	35
28	232	267	302	337	371	406	441	31	35
29	441	475	510	545	579	614	649	30	35
30	649	683	718	752	787	821	855	29	34
31	855	890	924	958	993	*027	*061	28	34
32	8.79 061	096	130	164	198	232	266	27	34
33	266	300	334	368	402	436	470	26	34
34	470	504	538	572	606	639	673	25	34
35	673	707	741	774	808	842	875	24	34
36	875	909	942	976	*009	*043	*076	23	34
37	8.80 076	110	143	177	210	243	277	22	34
38	277	310	343	376	409	443	476	21	33
39	476	509	542	575	608	641	674	20	33
40	674	707	740	773	806	839	872	19	33
41	872	905	937	970	*003	*036	*068	18	33
42	8.81 068	101	134	166	199	232	264	17	33
43	264	297	329	362	394	427	459	16	32
44	459	491	524	556	588	621	653	15	32
45	653	685	717	750	782	814	846	14	32
46	846	878	910	942	974	*006	*038	13	32
47	8.82 038	070	102	134	166	198	230	12	32
48	230	262	293	325	357	389	420	11	32
49	420	452	484	515	547	579	610	10	32
50	610	642	673	705	736	768	799	9	32
51	799	831	862	893	925	956	987	8	31
52	987	*019	*050	*081	*112	*144	*175	7	31
53	8.83 175	206	237	268	299	330	361	6	31
54	361	392	423	454	485	516	547	5	31
55	547	578	609	640	671	701	732	4	31
56	732	763	794	824	855	886	916	3	31
57	916	947	978	*008	*039	*069	*100	2	31
58	8.84 100	130	161	191	222	252	282	1	30
59	282	313	343	374	404	434	464	0	30
	60"	50"	40"	30"	20"	10"	0"	'	d.

P. P.

	41	40
1	4,1	4,0
2	8,2	8,0
3	12,3	12,0
4	16,4	16,0
5	20,5	20,0
6	24,6	24,0
7	28,7	28,0
8	32,8	32,0
9	36,9	36,0

	39	38
1	3,9	3,8
2	7,8	7,6
3	11,7	11,4
4	15,6	15,2
5	19,5	19,0
6	23,4	22,8
7	27,3	26,6
8	31,2	30,4
9	35,1	34,2

	37	36
1	3,7	3,6
2	7,4	7,2
3	11,1	10,8
4	14,8	14,4
5	18,5	18,0
6	22,2	21,6
7	25,9	25,2
8	29,6	28,8
9	33,3	32,4

	35	34
1	3,5	3,4
2	7,0	6,8
3	10,5	10,2
4	14,0	13,6
5	17,5	17,0
6	21,0	20,4
7	24,5	23,8
8	28,0	27,2
9	31,5	30,6

	33	32
1	3,3	3,2
2	6,6	6,4
3	9,9	9,6
4	13,2	12,8
5	16,5	16,0
6	19,8	19,2
7	23,1	22,4
8	26,4	25,6
9	29,7	28,8

	31	30
1	3,1	3,0
2	6,2	6,0
3	9,3	9,0
4	12,4	12,0
5	15,5	15,0
6	18,6	18,0
7	21,7	21,0
8	24,8	24,0
9	27,9	27,0

P. P.

86°   L. Cotg.

L. Cos. **L. Sin.** **4°**

9.99	'	0"	10"	20"	30"	40"	50"	60"		9.99	d.
894	0	8.84 858	889	419	449	479	509	589	59	898	30
893	1	539	569	599	629	659	688	718	58	892	30
892	2	718	748	778	808	838	867	897	57	891	30
891	3	897	927	957	986	*016	*045	*075	56	891	30
891	4	8.85 075	105	134	164	193	223	252	55	890	30
890	5	252	282	311	341	370	400	429	54	889	30
889	6	429	458	488	517	546	576	605	53	868	29
888	7	605	634	663	693	722	751	780	52	887	29
887	8	780	809	838	867	896	926	955	51	886	29
886	9	955	984	*013	*042	*070	*099	*128	50	885	29
885	10	8.86 128	157	186	215	244	278	301	49	884	29
884	11	301	330	359	388	416	445	474	48	883	29
883	12	474	502	531	560	588	617	645	47	882	29
882	13	645	674	703	731	760	788	816	46	881	29
881	14	816	845	873	902	930	958	987	45	880	28
880	15	987	*015	*048	*072	*100	*128	*156	44	879	28
879	16	8.87 156	185	218	241	269	297	325	43	879	28
879	17	325	354	382	410	438	466	494	42	878	28
878	18	494	522	550	578	606	634	661	41	877	28
877	19	661	689	717	745	778	801	829	40	876	28
876	20	829	856	884	912	940	967	995	39	875	28
875	21	995	*028	*050	*078	*106	*138	*161	38	874	28
874	22	8.88 161	188	216	248	271	298	326	37	873	28
873	23	326	358	381	408	436	463	490	36	872	27
872	24	490	518	545	572	600	627	654	35	871	27
871	25	654	681	709	786	768	790	817	34	870	27
870	26	817	845	872	899	926	958	980	33	869	27
869	27	980	*007	*084	*061	*088	*115	*142	32	868	27
868	28	8.89 142	169	196	223	250	277	304	31	867	27
867	29	304	330	357	384	411	438	464	30	866	27
866	30	464	491	518	545	571	598	625	29	865	27
865	31	625	651	678	704	731	758	784	28	864	26
864	32	784	811	837	864	890	917	943	27	863	26
863	33	943	970	996	*023	*049	*075	*102	26	862	26
862	34	8.90 102	128	154	181	207	233	260	25	861	26
861	35	260	286	312	338	364	391	417	24	860	26
860	36	417	443	469	495	521	548	574	23	859	26
859	37	574	600	626	652	678	704	730	22	858	26
858	38	730	756	782	808	834	859	885	21	857	26
857	39	885	911	937	963	989	*015	*040	20	856	26
856	40	8.91 040	066	092	118	148	169	195	19	855	26
855	41	195	221	246	272	298	323	349	18	854	26
854	42	349	374	400	426	451	477	502	17	853	26
853	43	502	528	553	579	604	630	655	16	852	26
852	44	655	680	706	731	757	782	807	15	851	25
851	45	807	833	858	888	909	984	959	14	850	25
850	46	959	984	*010	*085	*060	*085	*110	13	848	25
848	47	8.92 110	135	161	186	211	236	261	12	847	25
847	48	261	286	311	336	361	386	411	11	846	25
846	49	411	436	461	486	511	536	561	10	845	25
845	50	561	586	611	636	660	685	710	9	844	25
844	51	710	735	760	784	809	834	859	8	843	25
843	52	859	883	908	933	957	982	*007	7	842	25
842	53	8.93 007	031	056	081	105	130	154	6	841	24
841	54	154	179	203	228	253	277	301	5	840	24
840	55	301	326	350	375	399	424	448	4	839	24
839	56	448	472	497	521	546	570	594	3	838	24
838	57	594	619	643	667	691	716	740	2	837	24
837	58	740	764	788	812	837	861	885	1	836	24
836	59	885	909	933	957	981	*006	*030	0	834	24
		60"	50"	40"	30"	20"	10"	0"	'	9.99	d.

P. P.

	31	30
1	3,1	3,0
2	6,2	6,0
3	9,3	9,0
4	12,4	12,0
5	15,5	15,0
6	18,6	18,0
7	21,7	21,0
8	24,8	24,0
9	27,9	27,0

	29
1	2,9
2	5,8
3	8,7
4	11,6
5	14,5
6	17,4
7	20,3
8	23,2
9	26,1

	28	27
1	2,8	2,7
2	5,6	5,4
3	8,4	8,1
4	11,2	10,8
5	14,0	13,5
6	16,8	16,2
7	19,6	18,9
8	22,4	21,6
9	25,2	24,3

	26
1	2,6
2	5,2
3	7,8
4	10,4
5	13,0
6	15,6
7	18,2
8	20,8
9	23,4

	25	24
1	2,5	2,4
2	5,0	4,8
3	7,5	7,2
4	10,0	9,6
5	12,5	12,0
6	15,0	14,4
7	17,5	16,8
8	20,0	19,2
9	22,5	21,6

L. Tang.   4°

′	0″	10″	20″	30″	40″	50″	60″		d.	P. P.
0	8.84 464	495	525	555	585	615	646	59	30	
1	646	676	706	736	766	796	826	58	30	
2	826	856	886	916	946	976	*006	57	30	
3	8.85 006	036	065	095	125	155	185	56	30	
4	185	214	244	274	304	333	363	55	30	
5	363	392	422	452	481	511	540	54	30	
6	540	570	599	629	658	688	717	53	30	
7	717	747	776	805	835	864	893	52	29	
8	893	922	952	981	*010	*039	*069	51	29	
9	8.86 069	098	127	156	185	214	243	50	29	
10	243	272	301	330	359	388	417	49	29	
11	417	447	475	504	533	562	591	48	29	
12	591	619	648	677	706	734	763	47	29	
13	763	792	821	849	878	907	935	46	29	
14	935	964	992	*021	*049	*078	*106	45	28	
15	8.87 106	135	163	192	220	249	277	44	28	
16	277	305	334	362	390	419	447	43	28	
17	447	475	503	532	560	588	616	42	28	
18	616	644	673	701	729	757	785	41	28	
19	785	813	841	869	897	925	953	40	28	
20	953	981	*009	*037	*065	*092	*120	39	28	
21	8.88 120	148	176	204	231	259	287	38	28	
22	287	315	342	370	398	425	453	37	28	
23	453	481	508	536	563	591	618	36	28	
24	618	646	674	701	728	756	783	35	28	
25	783	811	838	866	893	920	948	34	28	
26	948	975	*002	*029	*057	*084	*111	33	27	
27	8.89 111	138	166	193	220	247	274	32	27	
28	274	301	328	355	383	410	437	31	27	
29	437	464	491	518	545	571	598	30	27	
30	598	625	652	679	706	733	760	29	27	
31	760	786	813	840	867	894	920	28	27	
32	920	947	974	*000	*027	*054	*080	27	27	
33	8.90 080	107	134	160	187	213	240	26	27	
34	240	266	293	319	346	372	399	25	26	
35	399	425	451	478	504	531	557	24	26	
36	557	583	610	636	662	688	715	23	26	
37	715	741	767	793	820	846	872	22	26	
38	872	898	924	950	976	*002	*029	21	26	
39	8.91 029	055	081	107	133	159	185	20	26	
40	185	211	236	262	288	314	340	19	26	
41	340	366	392	418	443	469	495	18	26	
42	495	521	547	572	598	624	650	17	26	
43	650	675	701	727	752	778	803	16	26	
44	803	829	855	880	906	931	957	15	26	
45	957	982	*008	*033	*059	*084	*110	14	26	
46	8.92 110	135	160	186	211	237	262	13	25	
47	262	287	313	338	363	388	414	12	25	
48	414	439	464	489	515	540	565	11	25	
49	565	590	615	640	665	691	716	10	25	
50	716	741	766	791	816	841	866	9	25	
51	866	891	916	941	966	991	*016	8	25	
52	8.93 016	040	065	090	115	140	165	7	25	
53	165	190	214	239	264	289	313	6	25	
54	313	338	363	388	412	437	462	5	25	
55	462	486	511	536	560	585	609	4	24	
56	609	634	658	683	707	732	756	3	24	
57	756	781	805	830	854	879	903	2	24	
58	903	928	952	976	*001	*025	*049	1	24	
59	8.94 049	074	098	122	147	171	195	0	24	
	60″	50″	40″	30″	20″	10″	0″	′	d.	P. P.

P. P.

	31	30
1	3,1	3,0
2	6,2	6,0
3	9,3	9,0
4	12,4	12,0
5	15,5	15,0
6	18,6	18,0
7	21,7	21,0
8	24,8	24,0
9	27,9	27,0

	29
1	2,9
2	5,8
3	8,7
4	11,6
5	14,5
6	17,4
7	20,3
8	23,2
9	26,1

	28	27
1	2,8	2,7
2	5,6	5,4
3	8,4	8,1
4	11,2	10,8
5	14,0	13,5
6	16,8	16,2
7	19,6	18,9
8	22,4	21,6
9	25,2	24,3

	26
1	2,6
2	5,2
3	7,8
4	10,4
5	13,0
6	15,6
7	18,2
8	20,8
9	23,4

	25	24
1	2,5	2,4
2	5,0	4,8
3	7,5	7,2
4	10,0	9,6
5	12,5	12,0
6	15,0	14,4
7	17,5	16,8
8	20,0	19,2
9	22,5	21,6

85°   L. Cotg.

9.99	'	0″	10″	20″	30″	40″	50″	60″		9.99	d.
884	0	8.94 080	054	078	102	126	150	174	59	888	24
888	1	174	198	222	246	270	294	317	58	882	24
882	2	317	341	365	389	418	437	461	57	881	24
881	3	− 461	484	508	532	556	580	608	56	880	24
880	4	608	627	651	675	698	722	746	55	829	24
829	5	746	769	798	817	840	864	887	54	828	24
826	6	887	911	935	958	982	*005	*029	53	827	24
827	7	8.95 029	052	076	099	123	146	170	52	825	24
825	8	170	193	216	240	263	287	310	51	824	23
824	9	310	333	357	380	403	427	450	50	823	23
823	10	450	473	496	520	543	566	589	49	822	23
822	11	589	613	636	659	682	705	728	48	821	23
821	12	728	752	775	798	821	844	867	47	820	23
820	13	867	890	913	936	959	982	*005	46	819	23
819	14	8.96 005	028	051	074	097	120	143	45	817	23
817	15	143	166	189	212	234	257	280	44	816	23
816	16	280	303	326	349	371	394	417	43	815	23
815	17	417	440	462	485	508	531	553	42	814	23
814	18	553	576	599	621	644	667	689	41	813	23
813	19	689	712	735	757	780	802	825	40	812	23
812	20	825	847	870	892	915	937	960	39	810	22
810	21	960	982	*005	*027	*050	*072	*095	38	809	22
809	22	8.97 095	117	139	162	184	207	229	37	808	22
808	23	229	251	274	296	318	341	363	36	807	22
807	24	363	385	407	430	452	474	496	35	806	22
806	25	496	518	541	563	585	607	629	34	804	22
804	26	629	651	674	696	718	740	762	33	803	22
803	27	762	784	806	828	850	872	894	32	802	22
802	28	894	916	938	960	982	*004	*026	31	801	22
801	29	8.98 026	048	070	092	114	135	157	30	800	22
800	30	157	179	201	223	245	266	288	29	798	22
798	31	288	310	332	354	375	397	419	28	797	22
797	32	419	441	462	484	506	527	549	27	796	22
796	33	549	571	592	614	636	657	679	26	795	22
795	34	679	701	722	744	765	787	808	25	793	22
793	35	808	830	851	873	894	916	937	24	792	22
792	36	937	959	980	*002	*023	*045	*066	23	791	22
791	37	8.99 066	087	109	130	152	173	194	22	790	21
790	38	194	216	237	258	280	301	322	21	788	21
788	39	322	343	365	386	407	428	450	20	787	21
787	40	450	471	492	513	534	556	577	19	786	21
786	41	577	598	619	640	661	682	704	18	785	21
785	42	704	725	746	767	788	809	830	17	783	21
783	43	830	851	872	893	914	935	956	16	782	21
782	44	956	977	998	*019	*040	*061	*082	15	781	21
781	45	9.00 082	103	123	144	165	186	207	14	780	21
780	46	207	228	249	269	290	311	332	13	778	21
778	47	332	353	373	394	415	436	456	12	777	21
777	48	456	477	498	518	539	560	581	11	776	21
776	49	581	601	622	642	663	684	704	10	775	21
775	50	704	725	746	766	787	807	828	9	773	21
773	51	828	848	869	889	910	930	951	8	772	20
772	52	951	971	992	*012	*033	*053	*074	7	771	20
771	53	9.01 074	094	115	135	155	176	196	6	769	20
769	54	196	217	237	257	278	298	318	5	768	20
768	55	318	339	359	379	399	420	440	4	767	20
767	56	440	460	480	501	521	541	561	3	765	20
765	57	561	582	602	622	642	662	682	2	764	20
764	58	682	702	722	742	762	783	803	1	763	20
763	59	803	823	843	863	883	903	923	0	761	20
		60″	50″	40″	30″	20″	10″	0″	'	9.99	d.

P. P.

24
1 | 2,4
2 | 4,8
3 | 7,2
4 | 9,6
5 | 12,0
6 | 14,4
7 | 16,8
8 | 19,2
9 | 21,6

23
1 | 2,8
2 | 4,6
3 | 6,9
4 | 9,2
5 | 11,5
6 | 13,8
7 | 16,1
8 | 18,4
9 | 20,7

22
1 | 2,2
2 | 4,4
3 | 6,6
4 | 8,8
5 | 11,0
6 | 13,2
7 | 15,4
8 | 17,6
9 | 19,8

21
1 | 2,1
2 | 4,2
3 | 6,3
4 | 8,4
5 | 10,5
6 | 12,6
7 | 14,7
8 | 16,8
9 | 18,9

20
1 | 2,0
2 | 4,0
3 | 6,0
4 | 8,0
5 | 10,0
6 | 12,0
7 | 14,0
8 | 16,0
9 | 18,0

L. Tang. 5°

′	0″	10″	20″	30″	40″	50″	60″		d.
0	8.94 195	219	244	268	292	316	340	59	24
1	340	365	389	418	437	461	485	58	24
2	485	509	533	557	581	606	630	57	24
3	630	654	678	702	725	749	773	56	24
4	773	797	821	845	869	893	917	55	24
5	917	941	964	988	*012	*036	*060	54	24
6	8.95 060	083	107	131	155	178	202	53	24
7	202	226	249	273	297	320	344	52	24
8	344	368	391	415	439	462	486	51	24
9	486	509	533	556	580	603	627	50	24
10	627	650	674	697	721	744	767	49	23
11	767	791	814	838	861	884	908	48	23
12	908	931	954	977	*001	*024	*047	47	23
13	8.96 047	071	094	117	140	163	187	46	23
14	187	210	233	256	279	302	325	45	23
15	325	349	372	395	418	441	464	44	23
16	464	487	510	533	556	579	602	43	23
17	602	625	648	671	694	717	739	42	23
18	739	762	785	808	831	854	877	41	23
19	877	899	922	945	968	991	*018	40	23
20	8.97 018	036	059	061	104	127	150	39	23
21	150	172	195	218	240	263	285	38	23
22	285	308	331	353	376	398	421	37	23
23	421	443	466	488	511	533	556	36	22
24	556	578	601	623	646	668	691	35	22
25	691	713	735	758	780	802	825	34	22
26	825	847	869	892	914	936	959	33	22
27	959	981	*003	*025	*048	*070	*092	32	22
28	8.98 092	114	136	159	181	203	225	31	22
29	225	247	269	291	314	336	358	30	22
30	358	380	402	424	446	468	490	29	22
31	490	512	534	556	578	600	622	28	22
32	622	644	666	687	709	731	753	27	22
33	753	775	797	819	841	862	884	26	22
34	884	906	928	950	971	993	*015	25	22
35	8.99 015	037	058	080	102	123	145	24	22
36	145	167	188	210	232	253	275	23	22
37	275	297	318	340	361	383	405	22	22
38	405	426	448	469	491	512	534	21	22
39	534	555	577	598	620	641	662	20	21
40	662	684	705	727	748	769	791	19	21
41	791	812	834	855	876	898	919	18	21
42	919	940	961	983	*004	*025	*046	17	21
43	9.00 046	068	089	110	131	153	174	16	21
44	174	195	216	237	258	280	301	15	21
45	301	322	343	364	385	406	427	14	21
46	427	448	469	490	511	532	553	13	21
47	553	574	595	616	637	658	679	12	21
48	679	700	721	742	763	784	805	11	21
49	805	826	846	867	888	909	930	10	21
50	930	951	971	992	*013	*034	*055	9	21
51	9.01 055	075	096	117	138	158	179	8	21
52	179	200	220	241	262	282	303	7	21
53	303	324	344	365	386	406	427	6	21
54	427	447	468	489	509	530	550	5	20
55	550	571	591	612	632	653	673	4	20
56	673	694	714	735	755	776	796	3	20
57	796	816	837	857	878	898	918	2	20
58	918	939	959	979	*000	*020	*040	1	20
59	9.02 040	061	081	101	121	142	162	0	20
	60″	50″	40″	30″	20″	10″	0″	′	d.

P. P.

25	
1	2,5
2	5,0
3	7,5
4	10,0
5	12,5
6	15,0
7	17,5
8	20,0
9	22,5

24	
1	2,4
2	4,8
3	7,2
4	9,6
5	12,0
6	14,4
7	16,8
8	19,2
9	21,6

23	
1	2,3
2	4,6
3	6,9
4	9,2
5	11,5
6	13,8
7	16,1
8	18,4
9	20,7

22	
1	2,2
2	4,4
3	6,6
4	8,8
5	11,0
6	13,2
7	15,4
8	17,6
9	19,8

21	
1	2,1
2	4,2
3	6,3
4	8,4
5	10,5
6	12,6
7	14,7
8	16,8
9	18,9

20	
1	2,0
2	4,0
3	6,0
4	8,0
5	10,0
6	12,0
7	14,0
8	16,0
9	18,0

84° L. Cotg.

'	L. Sin.	d.	L. Tang.	c.d.	L. Cotg.	L. Cos.	
0	9.01 923	120	9.02 162	121	0.97 888	9.99 761	60
1	9.02 048	120	9.02 283	121	0.97 717	9.99 760	59
2	9.02 168	120	9.02 404	121	0.97 596	9.99 759	58
3	9.02 288	119	9.02 525	121	0.97 475	9.99 757	57
4	9.02 402	118	9.02 645	120	0.97 355	9.99 756	56
5	9.02 520	119	9.02 766	121	0.97 234	9.99 755	55
6	9.02 689	118	9.02 885	119	0.97 115	9.99 753	54
7	9.02 757	117	9.03 005	120	0.96 995	9.99 752	53
8	9.02 874	118	9.03 124	119	0.96 876	9.99 751	52
9	9.02 992	117	9.03 242	118	0.96 758	9.99 749	51
10	9.03 109	117	9.03 361	119	0.96 639	9.99 748	50
11	9.03 226	116	9.03 479	118	0.96 521	9.99 747	49
12	9.03 342	116	9.03 597	118	0.96 403	9.99 745	48
13	9.03 458	116	9.03 714	117	0.96 286	9.99 744	47
14	9.03 574	116	9.03 832	118	0.96 168	9.99 742	46
15	9.03 690	115	9.03 948	116	0.96 052	9.99 741	45
16	9.03 805	115	9.04 065	117	0.95 935	9.99 740	44
17	9.03 920	115	9.04 181	116	0.95 819	9.99 738	43
18	9.04 034	114	9.04 297	116	0.95 703	9.99 737	42
19	9.04 149	115	9.04 413	116	0.95 587	9.99 736	41
20	9.04 262	113	9.04 528	115	0.95 472	9.99 734	40
21	9.04 376	114	9.04 643	115	0.95 357	9.99 733	39
22	9.04 490	114	9.04 758	115	0.95 242	9.99 731	38
23	9.04 603	113	9.04 873	115	0.95 127	9.99 730	37
24	9.04 715	112	9.04 987	114	0.95 013	9.99 728	36
25	9.04 828	113	9.05 101	114	0.94 899	9.99 727	35
26	9.04 940	112	9.05 214	113	0.94 786	9.99 726	34
27	9.05 052	112	9.05 328	114	0.94 672	9.99 724	33
28	9.05 164	112	9.05 441	113	0.94 559	9.99 723	32
29	9.05 275	111	9.05 553	112	0.94 447	9.99 721	31
30	9.05 386	111	9.05 666	113	0.94 334	9.99 720	30
31	9.05 497	111	9.05 778	112	0.94 222	9.99 718	29
32	9.05 607	110	9.05 890	112	0.94 110	9.99 717	28
33	9.05 717	110	9.06 002	112	0.93 998	9.99 716	27
34	9.05 827	110	9.06 118	111	0.93 887	9.99 714	26
35	9.05 937	110	9.06 224	111	0.93 776	9.99 713	25
36	9.06 046	109	9.06 335	111	0.93 665	9.99 711	24
37	9.06 155	109	9.06 445	110	0.93 555	9.99 710	23
38	9.06 264	109	9.06 556	111	0.93 444	9.99 708	22
39	9.06 372	108	9.06 666	110	0.93 334	9.99 707	21
40	9.06 481	109	9.06 775	109	0.93 225	9.99 705	20
41	9.06 589	108	9.06 885	110	0.93 115	9.99 704	19
42	9.06 696	107	9.06 994	109	0.93 006	9.99 702	18
43	9.06 804	108	9.07 103	109	0.92 897	9.99 701	17
44	9.06 911	107	9.07 211	108	0.92 789	9.99 699	16
45	9.07 018	107	9.07 320	109	0.92 680	9.99 698	15
46	9.07 124	106	9.07 428	108	0.92 572	9.99 696	14
47	9.07 231	107	9.07 536	108	0.92 464	9.99 695	13
48	9.07 337	106	9.07 643	107	0.92 357	9.99 693	12
49	9.07 442	105	9.07 751	108	0.92 249	9.99 692	11
50	9.07 548	106	9.07 858	107	0.92 142	9.99 690	10
51	9.07 653	105	9.07 964	106	0.92 086	9.99 689	9
52	9.07 758	105	9.08 071	107	0.91 929	9.99 687	8
53	9.07 863	105	9.08 177	106	0.91 823	9.99 686	7
54	9.07 968	105	9.08 283	106	0.91 717	9.99 684	6
55	9.08 072	104	9.08 389	106	0.91 611	9.99 683	5
56	9.08 176	104	9.08 495	105	0.91 505	9.99 681	4
57	9.08 280	104	9.08 600	105	0.91 400	9.99 680	3
58	9.08 383	103	9.08 705	105	0.91 295	9.99 678	2
59	9.08 486	103	9.08 810	104	0.91 190	9.99 677	1
60	9.08 589	103	9.08 914		0.91 086	9.99 675	0
	L. Cos.	d.	L. Cotg.	c.d.	L. Tang.	L. Sin.	'

P. P.

	121	120	119	118
1	2,0	2,0	2,0	2,0
2	4,0	4,0	4,0	3,9
3	6,0	6,0	6,0	5,9
4	8,1	8,0	7,9	7,9
5	10,1	10,0	9,9	9,8
6	12,1	12,0	11,9	11,8
7	14,1	14,0	13,9	13,8
8	16,1	16,0	15,9	15,7
9	18,2	18,0	17,8	17,7
10	20,2	20,0	19,8	19,7
20	40,3	40,0	39,7	39,3
30	60,5	60,0	59,5	59,0
40	80,7	80,0	79,3	78,7
50	100,8	100,0	99,2	98,3

	117	116	115	114
1	2,0	1,9	1,9	1,9
2	3,9	3,9	3,8	3,8
3	5,8	5,8	5,8	5,7
4	7,8	7,7	7,7	7,6
5	9,8	9,7	9,6	9,5
6	11,7	11,6	11,5	11,4
7	13,6	13,5	13,4	13,3
8	15,6	15,5	15,3	15,2
9	17,6	17,4	17,2	17,1
10	19,5	19,3	19,2	19,0
20	39,0	38,7	38,3	38,0
30	58,5	58,0	57,5	57,0
40	78,0	77,3	76,7	76,0
50	97,5	96,7	95,8	95,0

	113	112	111	110
1	1,9	1,9	1,8	1,8
2	3,8	3,7	3,7	3,7
3	5,6	5,6	5,6	5,5
4	7,5	7,5	7,4	7,3
5	9,4	9,3	9,2	9,2
6	11,3	11,2	11,1	11,0
7	13,2	13,1	13,0	12,8
8	15,1	14,9	14,8	14,7
9	17,0	16,8	16,6	16,5
10	18,8	18,7	18,5	18,3
20	37,7	37,3	37,0	36,7
30	56,5	56,0	55,5	55,0
40	75,3	74,7	74,0	73,3
50	94,2	93,3	92,5	91,7

	109	108	107	106
1	1,8	1,8	1,8	1,8
2	3,6	3,6	3,6	3,5
3	5,4	5,4	5,4	5,3
4	7,3	7,2	7,1	7,1
5	9,1	9,0	8,9	8,8
6	10,9	10,8	10,7	10,6
7	12,7	12,6	12,5	12,4
8	14,5	14,4	14,3	14,1
9	16,4	16,2	16,0	15,9
10	18,2	18,0	17,8	17,7
20	36,3	36,0	35,7	35,3
30	54,5	54,0	53,5	53,0
40	72,7	72,0	71,3	70,7
50	90,8	90,0	89,2	88,3

'	L. Sin.	d.	L. Tang.	c.d.	L. Cotg.	L. Cos.		P. P.
0	9.08 589	108	9.08 914	105	0.91 086	9.99 675	60	
1	9.08 692	108	9.09 019	104	0.90 981	9.99 674	59	**105 104 103 102**
2	9.08 795	102	9.09 123	104	0.90 877	9.99 672	58	1 1,8 1,7 1,7 1,7
3	9.08 897	102	9.09 227	104	0.90 778	9.99 670	57	2 3,5 3,5 3,4 3,4
4	9.08 999	102	9.09 330	103	0.90 670	9.99 669	56	3 5,2 5,2 5,1 5,1
								4 7,0 6,9 6,9 6,8
5	9.09 101	101	9.09 484	103	0.90 566	9.99 667	55	5 8,8 8,7 8,6 8,5
6	9.09 202	102	9.09 587	103	0.90 463	9.99 666	54	6 10,5 10,4 10,8 10,2
7	9.09 304	101	9.09 640	102	0.90 360	9.99 664	53	7 12,2 12,1 12,0 11,9
8	9.09 405	101	9.09 742	103	0.90 258	9.99 663	52	8 14,0 13,9 13,7 13,6
9	9.09 506	100	9.09 845	102	0.90 155	9.99 661	51	9 15,8 15,6 15,4 15,3
10	9.09 606	101	9.09 947	102	0.90 053	9.99 659	50	10 17,5 17,3 17,2 17,0
11	9.09 707	100	9.10 049	101	0.89 951	9.99 658	49	20 35,0 34,7 34,3 34,0
12	9.09 807	100	9.10 150	102	0.89 850	9.99 656	48	30 52,5 52,0 51,5 51,0
13	9.09 907	99	9.10 252	101	0.89 748	9.99 655	47	40 70,0 69,3 68,7 68,0
14	9.10 006	100	9.10 353	101	0.89 647	9.99 653	46	50 87,5 86,7 85,8 85,0
15	9.10 106	99	9.10 454	101	0.89 546	9.99 651	45	
16	9.10 205	99	9.10 555	101	0.89 445	9.99 650	44	**101 100 99 98**
17	9.10 304	98	9.10 656	100	0.89 344	9.99 648	43	1 1,7 1,7 1,6 1,6
18	9.10 402	99	9.10 756	100	0.89 244	9.99 647	42	2 3,4 3,3 3,3 3,3
19	9.10 501	98	9.10 856	100	0.89 144	9.99 645	41	3 5,0 5,0 5,0 4,9
20	9.10 599	98	9.10 956	100	0.89 044	9.99 643	40	4 6,7 6,7 6,6 6,5
21	9.10 697	98	9.11 056	99	0.88 944	9.99 642	39	5 8,4 8,3 8,2 8,2
22	9.10 795	98	9.11 155	99	0.88 845	9.99 640	38	6 10,1 10,0 9,9 9,8
23	9.10 893	97	9.11 254	99	0.88 746	9.99 638	37	7 11,8 11,7 11,6 11,4
24	9.10 990	97	9.11 353	99	0.88 647	9.99 637	36	8 13,5 13,8 13,2 13,1
								9 15,2 15,0 14,8 14,7
25	9.11 087	97	9.11 452	99	0.88 548	9.99 635	35	10 16,8 16,7 16,5 16,3
26	9.11 184	97	9.11 551	98	0.88 449	9.99 633	34	20 33,7 33,3 33,0 32,7
27	9.11 281	96	9.11 649	98	0.88 351	9.99 632	33	30 50,5 50,0 49,5 49,0
28	9.11 377	97	9.11 747	98	0.88 253	9.99 630	32	40 67,3 66,7 66,0 65,3
29	9.11 474	96	9.11 845	98	0.88 155	9.99 629	31	50 84,2 83,3 82,5 81,7
30	9.11 570	96	9.11 943	97	0.88 057	9.99 627	30	
31	9.11 666	95	9.12 040	98	0.87 960	9.99 625	29	**97 96 95 94**
32	9.11 761	96	9.12 138	97	0.87 862	9.99 624	28	
33	9.11 857	95	9.12 235	97	0.87 765	9.99 622	27	1 1,6 1,6 1,6 1,6
34	9.11 952	95	9.12 332	96	0.87 668	9.99 620	26	2 3,2 3,2 3,2 3,1
								3 4,8 4,8 4,8 4,7
35	9.12 047	95	9.12 428	97	0.87 572	9.99 618	25	4 6,5 6,4 6,3 6,3
36	9.12 142	94	9.12 525	96	0.87 475	9.99 617	24	5 8,1 8,0 7,9 7,8
37	9.12 236	95	9.12 621	96	0.87 379	9.99 615	23	6 9,7 9,6 9,5 9,4
38	9.12 331	94	9.12 717	96	0.87 283	9.99 613	22	7 11,3 11,2 11,1 11,0
39	9.12 425	94	9.12 813	96	0.87 187	9.99 612	21	8 12,9 12,8 12,7 12,5
								9 14,6 14,4 14,2 14,1
40	9.12 519	93	9.12 909	95	0.87 091	9.99 610	20	10 16,2 16,0 15,8 15,7
41	9.12 612	94	9.13 004	95	0.86 996	9.99 608	19	20 32,3 32,0 31,7 31,3
42	9.12 706	93	9.13 099	95	0.86 901	9.99 607	18	30 48,5 48,0 47,5 47,0
43	9.12 799	93	9.13 194	95	0.86 806	9.99 605	17	40 64,7 64,0 63,3 62,7
44	9.12 892	93	9.13 289	95	0.86 711	9.99 603	16	50 80,8 80,0 79,2 78,3
45	9.12 985	93	9.13 384	94	0.86 616	9.99 601	15	
46	9.13 078	93	9.13 478	95	0.86 522	9.99 600	14	**93 92 91 90**
47	9.13 171	92	9.13 573	94	0.86 427	9.99 598	13	1 1,6 1,5 1,5 1,5
48	9.13 263	92	9.13 667	94	0.86 333	9.99 596	12	2 3,1 3,1 3,0 3,0
49	9.13 355	92	9.13 761	93	0.86 239	9.99 595	11	3 4,6 4,6 4,6 4,5
								4 6,2 6,1 6,1 6,0
50	9.13 447	92	9.13 854	94	0.86 146	9.99 593	10	5 7,8 7,7 7,6 7,5
51	9.13 539	91	9.13 948	93	0.86 052	9.99 591	9	6 9,3 9,2 9,1 9,0
52	9.13 630	92	9.14 041	93	0.85 959	9.99 589	8	7 10,8 10,7 10,6 10,5
53	9.13 722	91	9.14 134	93	0.85 866	9.99 588	7	8 12,4 12,3 12,1 12,0
54	9.13 813	91	9.14 227	93	0.85 773	9.99 586	6	9 14,0 13,8 13,6 13,5
55	9.13 904	90	9.14 320	92	0.85 680	9.99 584	5	10 15,5 15,3 15,2 15,0
56	9.13 994	91	9.14 412	92	0.85 588	9.99 582	4	20 31,0 30,7 30,3 30,0
57	9.14 085	90	9.14 504	93	0.85 496	9.99 581	3	30 46,5 46,0 45,5 45,0
58	9.14 175	91	9.14 597	91	0.85 403	9.99 579	2	40 62,0 61,3 60,7 60,0
59	9.14 266	90	9.14 688	92	0.85 312	9.99 577	1	50 77,5 76,7 75,8 75,0
60	9.14 356		9.14 780		0.85 220	9.99 575	0	
	L. Cos.	d.	L. Cotg.	c.d.	L. Tang.	L. Sin.	'	P. P.

8°

'	L. Sin.	d.	L. Tang.	c.d.	L. Cotg.	L. Cos.	
0	9.14 356	89	9.14 780	92	0.85 220	9.99 575	60
1	9.14 445	90	9.14 872	91	0.85 128	9.99 574	59
2	9.14 535	89	9.14 963	91	0.85 037	9.99 572	58
3	9.14 624	90	9.15 054	91	0.84 946	9.99 570	57
4	9.14 714	89	9.15 145	91	0.84 855	9.99 568	56
5	9.14 803	88	9.15 236	91	0.84 764	9.99 566	55
6	9.14 891	89	9.15 327	90	0.84 673	9.99 565	54
7	9.14 980	89	9.15 417	91	0.84 583	9.99 563	53
8	9.15 069	88	9.15 508	90	0.84 492	9.99 561	52
9	9.15 157	88	9.15 598	90	0.84 402	9.99 559	51
10	9.15 245	88	9.15 688	89	0.84 312	9.99 557	50
11	9.15 333	88	9.15 777	90	0.84 223	9.99 556	49
12	9.15 421	87	9.15 867	89	0.84 133	9.99 554	48
13	9.15 508	88	9.15 956	90	0.84 044	9.99 552	47
14	9.15 596	87	9.16 046	89	0.83 954	9.99 550	46
15	9.15 683	87	9.16 135	89	0.83 865	9.99 548	45
16	9.15 770	87	9.16 224	88	0.83 776	9.99 546	44
17	9.15 857	87	9.16 312	89	0.83 688	9.99 545	43
18	9.15 944	86	9.16 401	88	0.83 599	9.99 543	42
19	9.16 030	86	9.16 489	88	0.83 511	9.99 541	41
20	9.16 116	87	9.16 577	88	0.83 423	9.99 539	40
21	9.16 203	86	9.16 665	88	0.83 335	9.99 537	39
22	9.16 289	85	9.16 753	88	0.83 247	9.99 535	38
23	9.16 374	86	9.16 841	87	0.83 159	9.99 533	37
24	9.16 460	85	9.16 928	88	0.83 072	9.99 532	36
25	9.16 545	86	9.17 016	87	0.82 984	9.99 530	35
26	9.16 631	85	9.17 108	87	0.82 897	9.99 528	34
27	9.16 716	85	9.17 190	87	0.82 810	9.99 526	33
28	9.16 801	85	9.17 277	86	0.82 723	9.99 524	32
29	9.16 886	84	9.17 363	87	0.82 637	9.99 522	31
30	9.16 970	85	9.17 450	86	0.82 550	9.99 520	30
31	9.17 055	84	9.17 536	86	0.82 464	9.99 518	29
32	9.17 139	84	9.17 622	86	0.82 378	9.99 517	28
33	9.17 223	84	9.17 708	86	0.82 292	9.99 515	27
34	9.17 307	84	9.17 794	86	0.82 206	9.99 513	26
35	9.17 391	83	9.17 880	85	0.82 120	9.99 511	25
36	9.17 474	84	9.17 965	86	0.82 035	9.99 509	24
37	9.17 558	83	9.18 051	85	0.81 949	9.99 507	23
38	9.17 641	83	9.18 136	85	0.81 864	9.99 505	22
39	9.17 724	83	9.18 221	85	0.81 779	9.99 503	21
40	9.17 807	83	9.18 306	85	0.81 694	9.99 501	20
41	9.17 890	83	9.18 391	84	0.81 609	9.99 499	19
42	9.17 973	82	9.18 475	85	0.81 525	9.99 497	18
43	9.18 055	82	9.18 560	84	0.81 440	9.99 495	17
44	9.18 137	83	9.18 644	84	0.81 356	9.99 494	16
45	9.18 220	82	9.18 728	84	0.81 272	9.99 492	15
46	9.18 302	81	9.18 812	84	0.81 188	9.99 490	14
47	9.18 383	82	9.18 896	83	0.81 104	9.99 488	13
48	9.18 465	82	9.18 979	84	0.81 021	9.99 486	12
49	9.18 547	81	9.19 063	83	0.80 937	9.99 484	11
50	9.18 628	81	9.19 146	83	0.80 854	9.99 482	10
51	9.18 709	81	9.19 229	83	0.80 771	9.99 480	9
52	9.18 790	81	9.19 312	83	0.80 688	9.99 478	8
53	9.18 871	81	9.19 395	83	0.80 605	9.99 476	7
54	9.18 952	81	9.19 478	83	0.80 522	9.99 474	6
55	9.19 033	80	9.19 561	82	0.80 439	9.99 472	5
56	9.19 113	80	9.19 643	82	0.80 357	9.99 470	4
57	9.19 193	80	9.19 725	82	0.80 275	9.99 468	3
58	9.19 273	80	9.19 807	82	0.80 193	9.99 466	2
59	9.19 353	80	9.19 889	82	0.80 111	9.99 464	1
60	9.19 433		9.19 971		0.80 029	9.99 462	0
	L. Cos.	d.	L. Cotg.	c.d.	L. Tang.	L. Sin.	'

P. P.

	92	91	90
1	1,5	1,5	1,5
2	3,1	3,0	3,0
3	4,6	4,6	4,5
4	6,1	6,1	6,0
5	7,7	7,6	7,5
6	9,2	9,1	9,0
7	10,7	10,6	10,5
8	12,3	12,1	12,0
9	13,8	13,6	13,5
10	15,3	15,2	15,0
20	30,7	30,3	30,0
30	46,0	45,5	45,0
40	61,3	60,7	60,0
50	76,7	75,8	75,0

	89	88	87
1	1,5	1,5	1,4
2	3,0	2,9	2,9
3	4,4	4,4	4,4
4	5,9	5,9	5,8
5	7,4	7,3	7,2
6	8,9	8,8	8,7
7	10,4	10,3	10,2
8	11,9	11,7	11,6
9	13,4	13,2	13,0
10	14,8	14,7	14,5
20	29,7	29,3	29,0
30	44,5	44,0	43,5
40	59,3	58,7	58,0
50	74,2	73,3	72,5

	86	85	84
1	1,4	1,4	1,4
2	2,9	2,8	2,8
3	4,3	4,2	4,2
4	5,7	5,7	5,6
5	7,2	7,1	7,0
6	8,6	8,5	8,4
7	10,0	9,9	9,8
8	11,5	11,3	11,2
9	12,9	12,8	12,6
10	14,3	14,2	14,0
20	28,7	28,3	28,0
30	43,0	42,5	42,0
40	57,3	56,7	56,0
50	71,7	70,8	70,0

	83	82	81
1	1,4	1,4	1,4
2	2,8	2,7	2,7
3	4,2	4,1	4,0
4	5,5	5,5	5,4
5	6,9	6,8	6,8
6	8,3	8,2	8,1
7	9,7	9,6	9,4
8	11,1	10,9	10,8
9	12,4	12,3	12,2
10	13,8	13,7	13,5
20	27,7	27,3	27,0
30	41,5	41,0	40,5
40	55,3	54,7	54,0
50	69,2	68,3	67,5

P. P.

9°

'	L. Sin.	d.	L. Tang.	c.d.	L. Cotg.	L. Cos.	
0	9.19 488	80	9.19 971	82	0.80 029	9.99 462	60
1	9.19 518	79	9.20 058	81	0.79 947	9.99 460	59
2	9.19 592	80	9.20 184	82	0.79 866	9.99 458	58
3	9.19 672	79	9.20 216	81	0.79 784	9.99 456	57
4	9.19 751	79	9.20 297	81	0.79 708	9.99 454	56
5	9.19 880	79	9.20 378	81	0.79 622	9.99 452	55
6	9.19 909	79	9.20 459	81	0.79 541	9.99 450	54
7	9.19 988	79	9.20 540	81	0.79 460	9.99 448	53
8	9.20 067	78	9.20 621	80	0.79 379	9.99 446	52
9	9.20 145	78	9.20 701	81	0.79 299	9.99 444	51
10	9.20 223	79	9.20 782	80	0.79 218	9.99 442	50
11	9.20 302	78	9.20 862	80	0.79 188	9.99 440	49
12	9.20 380	78	9.20 942	80	0.79 058	9.99 438	48
13	9.20 458	77	9.21 022	80	0.78 978	9.99 436	47
14	9.20 535	78	9.21 102	80	0.78 898	9.99 434	46
15	9.20 613	78	9.21 182	79	0.78 818	9.99 432	45
16	9.20 691	77	9.21 261	80	0.78 789	9.99 429	44
17	9.20 768	77	9.21 341	79	0.78 659	9.99 427	43
18	9.20 845	77	9.21 420	79	0.78 580	9.99 425	42
19	9.20 922	77	9.21 499	79	0.78 501	9.99 423	41
20	9.20 999	77	9.21 578	79	0.78 422	9.99 421	40
21	9.21 076	77	9.21 657	79	0.78 343	9.99 419	39
22	9.21 153	76	9.21 786	78	0.78 264	9.99 417	38
23	9.21 229	77	9.21 814	79	0.78 186	9.99 415	37
24	9.21 306	76	9.21 893	78	0.78 107	9.99 413	36
25	9.21 382	76	9.21 971	78	0.78 029	9.99 411	35
26	9.21 458	76	9.22 049	78	0.77 951	9.99 409	34
27	9.21 584	76	9.22 127	78	0.77 873	9.99 407	33
28	9.21 610	75	9.22 205	78	0.77 795	9.99 404	32
29	9.21 685	76	9.22 283	78	0.77 717	9.99 402	31
30	9.21 761	75	9.22 361	77	0.77 639	9.99 400	30
31	9.21 836	76	9.22 438	78	0.77 562	9.99 398	29
32	9.21 912	75	9.22 516	77	0.77 484	9.99 396	28
33	9.21 987	75	9.22 593	77	0.77 407	9.99 394	27
34	9.22 062	75	9.22 670	77	0.77 330	9.99 392	26
35	9.22 137	74	9.22 747	77	0.77 253	9.99 390	25
36	9.22 211	75	9.22 824	77	0.77 176	9.99 388	24
37	9.22 286	75	9.22 901	76	0.77 099	9.99 385	23
38	9.22 361	74	9.22 977	77	0.77 023	9.99 383	22
39	9.22 485	74	9.23 054	76	0.76 946	9.99 381	21
40	9.22 509	74	9.23 180	76	0.76 870	9.99 379	20
41	9.22 583	74	9.23 206	77	0.76 794	9.99 377	19
42	9.22 657	74	9.23 283	76	0.76 717	9.99 375	18
43	9.22 731	74	9.23 359	76	0.76 641	9.99 372	17
44	9.22 805	73	9.23 435	75	0.76 565	9.99 370	16
45	9.22 878	74	9.23 510	76	0.76 490	9.99 368	15
46	9.22 952	73	9.23 586	75	0.76 414	9.99 366	14
47	9.23 025	73	9.23 661	76	0.76 339	9.99 364	13
48	9.23 098	73	9.23 737	75	0.76 263	9.99 362	12
49	9.23 171	73	9.23 812	75	0.76 188	9.99 359	11
50	9.23 244	73	9.23 887	75	0.76 113	9.99 357	10
51	9.23 317	73	9.23 962	75	0.76 088	9.99 355	9
52	9.23 390	72	9.24 037	75	0.75 963	9.99 353	8
53	9.23 462	73	9.24 112	74	0.75 888	9.99 351	7
54	9.23 535	72	9.24 186	75	0.75 814	9.99 348	6
55	9.23 607	72	9.24 261	74	0.75 739	9.99 346	5
56	9.23 679	73	9.24 335	75	0.75 665	9.99 344	4
57	9.23 752	71	·9.24 410	74	0.75 590	9.99 342	3
58	9.23 823	72	9.24 484	74	0.75 516	9.99 340	2
59	9.23 895	72	9.24 558	74	0.75 442	9.99 337	1
60	9.23 967		9.24 632		0.75 368	9.99 335	0
	L. Cos.	d.	L. Cotg.	c.d.	L. Tang.	L. Sin.	'

P.P.

	80	79	78	77
1	1,3	1,3	1,3	1,3
2	2,7	2,6	2,6	2,6
3	4,0	4,0	3,9	3,8
4	5,3	5,3	5,2	5,1
5	6,7	6,6	6,5	6,4
6	8,0	7,9	7,8	7,7
7	9,3	9,2	9,1	9,0
8	10,7	10,5	10,4	10,3
9	12,0	11,8	11,7	11,6
10	13,3	13,2	13,0	12,8
20	26,7	26,3	26,0	25,7
30	40,0	39,5	39,0	38,5
40	53,3	52,7	52,0	51,3
50	66,7	65,8	65,0	64,2

	76	75	74	73
1	1,3	1,2	1,2	1,2
2	2,5	2,5	2,5	2,4
3	3,8	3,8	3,7	3,6
4	5,1	5,0	4,9	4,9
5	6,3	6,2	6,2	6,1
6	7,6	7,5	7,4	7,3
7	8,9	8,8	8,6	8,5
8	10,1	10,0	9,9	9,7
9	11,4	11,2	11,1	11,0
10	12,7	12,5	12,3	12,2
20	25,3	25,0	24,7	24,3
30	38,0	37,5	37,0	36,5
40	50,7	50,0	49,3	48,7
50	63,3	62,5	61,7	60,8

	72	71	3	2
1	1,2	1,2	0,0	0,0
2	2,4	2,4	0,1	0,1
3	3,6	3,6	0,2	0,1
4	4,8	4,7	0,2	0,1
5	6,0	5,9	0,2	0,2
6	7,2	7,1	0,3	0,2
7	8,4	8,3	0,4	0,2
8	9,6	9,5	0,4	0,3
9	10,8	10,6	0,4	0,3
10	12,0	11,8	0,5	0,3
20	24,0	23,7	1,0	0,7
30	36,0	35,5	1,5	1,0
40	48,0	47,3	2,0	1,3
50	60,0	59,2	2,5	1,7

| | | | | | | | | P. P. |

80°

′	L. Sin.	d.	L. Tang.	c.d.	L. Cotg.	L. Cos.	d.		P. P.		
0	9.23 967	72	9.24 632	74	0.75 368	9.99 335	2	60			
1	9.24 039	71	9.24 706	73	0.75 294	9.99 333	2	59			
2	9.24 110	71	9.24 779	74	0.75 221	9.99 331	3	58			
3	9.24 181	72	9.24 853	73	0.75 147	9.99 328	2	57			
4	9.24 253	71	9.24 926	74	0.75 074	9.99 326	2	56	**74**	**73**	**72**
5	9.24 324	71	9.25 000	73	0.75 000	9.99 324	2	55	1 1,2	1,2	1,2
6	9.24 395	71	9.25 073	73	0.74 927	9.99 322	3	54	2 2,5	2,4	2,4
7	9.24 466	70	9.25 146	73	0.74 854	9.99 319	2	53	3 3,7	3,6	3,6
8	9.24 536	71	9.25 219	73	0.74 781	9.99 317	2	52	4 4,9	4,9	4,8
9	9.24 607	70	9.25 292	73	0.74 708	9.99 315	2	51			
10	9.24 677	71	9.25 365	72	0.74 635	9.99 313	3	50	5 6,2	6,1	6,0
11	9.24 748	70	9.25 437	73	0.74 563	9.99 310	2	49	6 7,4	7,3	7,2
12	9.24 818	70	9.25 510	72	0.74 490	9.99 308	2	48	7 8,6	8,5	8,4
13	9.24 888	70	9.25 582	73	0.74 418	9.99 306	2	47	8 9,9	9,7	9,6
14	9.24 958	70	9.25 655	72	0.74 345	9.99 304	3	46	9 11,1	11,0	10,8
15	9.25 028	70	9.25 727	72	0.74 273	9.99 301	2	45	10 12,3	12,2	12,0
16	9.25 098	70	9.25 799	72	0.74 201	9.99 299	2	44	20 24,7	24,3	24,0
17	9.25 168	69	9.25 871	72	0.74 129	9.99 297	3	43	30 37,0	36,5	36,0
18	9.25 237	70	9.25 943	72	0.74 057	9.99 294	2	42	40 49,3	48,7	48,0
19	9.25 307	69	9.26 015	71	0.73 985	9.99 292	2	41	50 61,7	60,8	60,0
20	9.25 376	69	9.26 086	72	0.73 914	9.99 290	2	40			
21	9.25 445	69	9.26 158	71	0.73 842	9.99 288	3	39			
22	9.25 514	69	9.26 229	72	0.73 771	9.99 285	2	38			
23	9.25 583	69	9.26 301	71	0.73 699	9.99 283	2	37	**71**	**70**	**69**
24	9.25 652	69	9.26 372	71	0.73 628	9.99 281	3	36	1 1,2	1,2	1,2
25	9.25 721	69	9.26 443	71	0.73 557	9.99 278	2	35	2 2,4	2,3	2,3
26	9.25 790	68	9.26 514	71	0.73 486	9.99 276	2	34	3 3,6	3,5	3,4
27	9.25 858	69	9.26 585	70	0.73 415	9.99 274	3	33	4 4,7	4,7	4,6
28	9.25 927	68	9.26 655	71	0.73 345	9.99 271	2	32			
29	9.25 995	68	9.26 726	71	0.73 274	9.99 269	2	31	5 5,9	5,8	5,8
30	9.26 063	68	9.26 797	70	0.73 203	9.99 267	3	30	6 7,1	7,0	6,9
31	9.26 131	68	9.26 867	70	0.73 133	9.99 264	2	29	7 8,3	8,2	8,0
32	9.26 199	68	9.26 937	71	0.73 063	9.99 262	2	28	8 9,5	9,3	9,2
33	9.26 267	68	9.27 008	70	0.72 992	9.99 260	3	27	9 10,6	10,5	10,4
34	9.26 335	68	9.27 078	70	0.72 922	9.99 257	2	26			
35	9.26 403	67	9.27 148	70	0.72 852	9.99 255	3	25	10 11,8	11,7	11,5
36	9.26 470	68	9.27 218	70	0.72 782	9.99 252	2	24	20 23,7	23,3	23,0
37	9.26 538	67	9.27 288	69	0.72 712	9.99 250	2	23	30 35,5	35,0	34,5
38	9.26 605	67	9.27 357	70	0.72 643	9.99 248	3	22	40 47,3	46,7	46,0
39	9.26 672	67	9.27 427	69	0.72 573	9.99 245	2	21	50 59,2	58,3	57,5
40	9.26 739	67	9.27 496	70	0.72 504	9.99 243	2	20			
41	9.26 806	67	9.27 566	69	0.72 434	9.99 241	3	19			
42	9.26 873	67	9.27 635	69	0.72 365	9.99 238	3	18			
43	9.26 940	67	9.27 704	69	0.72 296	9.99 236	3	17	**68**	**67**	**66**
44	9.27 007	66	9.27 773	69	0.72 227	9.99 233	2	16	1 1,1	1,1	1,1
45	9.27 073	67	9.27 842	69	0.72 158	9.99 231	2	15	2 2,3	2,2	2,2
46	9.27 140	66	9.27 911	69	0.72 089	9.99 229	3	14	3 3,4	3,4	3,3
47	9.27 206	67	9.27 980	69	0.72 020	9.99 226	2	13	4 4,5	4,5	4,4
48	9.27 273	66	9.28 049	68	0.71 951	9.99 224	3	12			
49	9.27 339	66	9.28 117	69	0.71 883	9.99 221	2	11	5 5,7	5,6	5,5
50	9.27 405	66	9.28 186	68	0.71 814	9.99 219	2	10	6 6,8	6,7	6,6
51	9.27 471	66	9.28 254	69	0.71 746	9.99 217	3	9	7 7,9	7,8	7,7
52	9.27 537	65	9.28 323	68	0.71 677	9.99 214	2	8	8 9,1	8,9	8,8
53	9.27 602	66	9.28 391	68	0.71 609	9.99 212	3	7	9 10,2	10,0	9,9
54	9.27 668	66	9.28 459	68	0.71 541	9.99 209	2	6			
55	9.27 734	65	9.28 527	68	0.71 473	9.99 207	3	5	10 11,3	11,2	11,0
56	9.27 799	65	9.28 595	67	0.71 405	9.99 204	2	4	20 22,7	22,3	22,0
57	9.27 864	66	9.28 662	68	0.71 338	9.99 202	3	3	30 34,0	33,5	33,0
58	9.27 930	65	9.28 730	68	0.71 270	9.99 200	3	2	40 45,3	44,7	44,0
59	9.27 995	65	9.28 798	67	0.71 202	9.99 197	2	1	50 56,7	55,8	55,0
60	9.28 060		9.28 865		0.71 135	9.99 195		0			

| | L. Cos. | d. | L. Cotg. | c.d. | L. Tang. | L. Sin. | d. | ′ | | P. P. | |

′	L. Sin.	d.	L. Tang.	c.d.	L. Cotg.	L. Cos.	d.		P. P.		
0	9.28 060	65	9.28 865	68	0.71 135	9.99 195	3	60			
1	9.28 125	65	9.28 933	67	0.71 067	9.99 192	2	59			
2	9.28 190	64	9.29 000	67	0.71 000	9.99 190	3	58			
3	9.28 254	65	9.29 067	67	0.70 933	9.99 187	2	57			
4	9.28 319	65	9.29 134	67	0.70 866	9.99 185	3	56		**65**	**64**	**63**
5	9.28 384	64	9.29 201	67	0.70 799	9.99 182	2	55	1	1,1	1,1	1.0
6	9.28 448	64	9.29 268	67	0.70 732	9.99 180	3	54	2	2,2	2,1	2.1
7	9.28 512	65	9.29 335	67	0.70 665	9.99 177	2	53	3	3,2	3,2	3.2
8	9.28 577	64	9.29 402	66	0.70 598	9.99 175	3	52	4	4,3	4,3	4 2
9	9.28 641	64	9.29 468	67	0.70 532	9.99 172	2	51				
10	9.28 705	64	9.29 535	66	0.70 465	9.99 170	3	50	5	5,4	5,3	5.2
11	9.28 769	64	9.29 601	67	0.70 399	9.99 167	2	49	6	6,5	6,4	6,3
12	9.28 833	63	9.29 668	66	0.70 332	9.99 165	3	48	7	7,6	7,5	7,4
13	9.28 896	64	9.29 734	66	0.70 266	9.99 162	2	47	8	8,7	8,5	8,4
14	9.28 960	64	9.29 800	66	0.70 200	9.99 160	3	46	9	9,8	9,6	9,4
15	9.29 024	63	9.29 866	66	0.70 134	9.99 157	2	45	10	10,8	10,7	10,5
16	9.29 087	63	9.29 932	66	0.70 068	9.99 155	3	44	20	21,7	21,3	21,0
17	9.29 150	64	9.29 998	66	0.70 002	9.99 152	3	43	30	32,5	32,0	31,5
18	9.29 214	63	9.30 064	66	0.69 936	9.99 150	3	42	40	43,3	42,7	42,0
19	9.29 277	63	9.30 130	65	0.69 870	9.99 147	2	41	50	54,2	53,3	52,5
20	9.29 340	68	9.30 195	66	0.69 805	9.99 145	3	40				
21	9.29 403	63	9.30 261	65	0.69 739	9.99 142	2	39				
22	9.29 466	63	9.30 326	65	0.69 674	9.99 140	3	38				
23	9.29 529	62	9.30 391	66	0.69 609	9.99 137	2	37				
24	9.29 591	63	9.30 457	65	0.69 543	9.99 135	3	36		**62**	**61**	**60**
25	9.29 654	62	9.30 522	65	0.69 478	9.99 132	3	35	1	1,0	1,0	1,0
26	9.29 716	63	9.30 587	65	0.69 413	9.99 130	3	34	2	2,1	2,0	2,0
27	9.29 779	62	9.30 652	65	0.69 348	9.99 127	3	33	3	3,1	3,0	3,0
28	9.29 841	62	9.30 717	65	0.69 283	9.99 124	2	32	4	4,1	4,1	4,0
29	9.29 903	63	9.30 782	64	0.69 218	9.99 122	3	31	5	5,2	5,1	5,0
30	9.29 966	62	9.30 846	65	0.69 154	9.99 119	2	30	6	6,2	6,1	6,0
31	9.30 028	62	9.30 911	64	0.69 089	9.99 117	3	29	7	7,2	7,1	7,0
32	9.30 090	61	9.30 975	65	0.69 025	9.99 114	2	28	8	8,3	8,1	8,0
33	9.30 151	62	9.31 040	64	0.68 960	9.99 112	3	27	9	9,3	9,2	9,0
34	9.30 213	62	9.31 104	64	0.68 896	9.99 109	3	26				
35	9.30 275	61	9.31 168	65	0.68 832	9.99 106	2	25	10	10,3	10,2	10,0
36	9.30 336	62	9.31 233	64	0.68 767	9.99 104	3	24	20	20,7	20,3	20,0
37	9.30 398	61	9.31 297	64	0.68 703	9.99 101	3	23	30	31,0	30,5	30,0
38	9.30 459	62	9.31 361	64	0.68 639	9.99 099	3	22	40	41,3	40,7	40,0
39	9.30 521	61	9.31 425	64	0.68 575	9.99 096	3	21	50	51,7	50,8	50,0
40	9.30 582	61	9.31 489	63	0.68 511	9.99 093	2	20				
41	9.30 643	61	9.31 552	64	0.68 448	9.99 091	3	19				
42	9.30 704	61	9.31 616	63	0.68 384	9.99 088	3	18				
43	9.30 765	61	9.31 679	64	0.68 321	9.99 086	3	17		**59**	**3**	**2**
44	9.30 826	61	9.31 743	63	0.68 257	9.99 083	3	16	1	1,0	0,0	0,0
45	9.30 887	60	9.31 806	64	0.68 194	9.99 080	2	15	2	2,0	0,1	0,1
46	9.30 947	61	9.31 870	63	0.68 130	9.99 078	3	14	3	3,0	0,2	0,1
47	9.31 008	60	9.31 933	63	0.68 067	9.99 075	3	13	4	3,9	0,2	0,1
48	9.31 068	61	9.31 996	63	0.68 004	9.99 072	2	12				
49	9.31 129	60	9.32 059	63	0.67 941	9.99 070	3	11	5	4,9	0,2	0,2
50	9.31 189	61	9.32 122	63	0.67 878	9.99 067	3	10	6	5,9	0,3	0,2
51	9.31 250	60	9.32 185	63	0.67 815	9.99 064	2	9	7	6,9	0,4	0,2
52	9.31 310	60	9.32 248	63	0.67 752	9.99 062	3	8	8	7,9	0,4	0,3
53	9.31 370	60	9.32 311	62	0.67 689	9.99 059	3	7	9	8,8	0,4	0,3
54	9.31 430	60	9.32 373	63	0.67 627	9.99 056	2	6				
55	9.31 490	59	9.32 436	62	0.67 564	9.99 054	3	5	10	9,8	0,5	0,3
56	9.31 549	60	9.32 498	63	0.67 502	9.99 051	3	4	20	19,7	1,0	0,7
57	9.31 609	60	9.32 561	62	0.67 439	9.99 048	3	3	30	29,5	1,5	1,0
58	9.31 669	59	9.32 623	62	0.67 377	9.99 046	2	2	40	39,3	2,0	1,3
59	9.31 728	60	9.32 685	62	0.67 315	9.99 043	3	1	50	49,2	2,5	1,7
60	9.31 788		9.32 747		0.67 253	9.99 040		0				
	L. Cos.	d.	L. Cotg.	c.d.	L. Tang.	L. Sin.	d.	′		P. P.		

12°

'	L. Sin.	d.	L. Tang.	c.d.	L. Cotg.	L. Cos.	d.	'
0	9.81 788	59	9.82 747	63	0.17 253	9.99 040	2	60
1	9.81 847	60	9.82 810	62	0.17 190	9.99 038	3	59
2	9.81 907	59	9.82 872	61	0.17 128	9.99 035	3	58
3	9.81 966	59	9.82 933	62	0.17 067	9.99 033	3	57
4	9.82 025	59	9.82 995	62	0.17 005	9.99 030	3	56
5	9.82 084	59	9.83 057	62	0.16 943	9.99 027	3	55
6	9.82 143	59	9.83 119	61	0.16 881	9.99 024	2	54
7	9.82 202	59	9.83 180	62	0.16 820	9.99 022	3	53
8	9.82 261	58	9.83 242	61	0.16 758	9.99 019	3	52
9	9.82 319	59	9.83 303	62	0.16 697	9.99 016	3	51
10	9.82 378	59	9.83 365	61	0.16 635	9.99 013	2	50
11	9.82 437	58	9.83 426	61	0.16 574	9.99 011	3	49
12	9.82 495	58	9.83 487	61	0.16 513	9.99 008	3	48
13	9.82 553	59	9.83 548	61	0.16 452	9.99 005	3	47
14	9.82 612	58	9.83 609	61	0.16 391	9.99 002	2	46
15	9.82 670	58	9.83 670	61	0.16 330	9.99 000	3	45
16	9.82 728	58	9.83 731	61	0.16 269	9.98 997	3	44
17	9.82 786	58	9.83 792	61	0.16 208	9.98 994	3	43
18	9.82 844	58	9.83 853	60	0.16 147	9.98 991	2	42
19	9.82 902	58	9.83 913	61	0.16 087	9.98 989	3	41
20	9.82 960	58	9.83 974	60	0.16 026	9.98 986	3	40
21	9.83 018	57	9.84 034	61	0.15 966	9.98 983	3	39
22	9.83 075	58	9.84 095	60	0.15 905	9.98 980	2	38
23	9.83 133	57	9.84 155	60	0.15 845	9.98 978	3	37
24	9.83 190	58	9.84 215	61	0.15 785	9.98 975	3	36
25	9.83 248	57	9.84 276	60	0.15 724	9.98 972	3	35
26	9.83 305	57	9.84 336	60	0.15 664	9.98 969	2	34
27	9.83 362	58	9.84 396	60	0.15 604	9.98 967	3	33
28	9.83 420	57	9.84 456	60	0.15 544	9.98 964	3	32
29	9.83 477	57	9.84 516	60	0.15 484	9.98 961	3	31
30	9.83 534	57	9.84 576	59	0.15 424	9.98 958	3	30
31	9.83 591	56	9.84 635	60	0.15 365	9.98 955	2	29
32	9.83 647	57	9.84 695	60	0.15 305	9.98 953	3	28
33	9.83 704	57	9.84 755	59	0.15 245	9.98 950	3	27
34	9.83 761	57	9.84 814	60	0.15 186	9.98 947	3	26
35	9.83 818	56	9.84 874	59	0.15 126	9.98 944	3	25
36	9.83 874	57	9.84 933	59	0.15 067	9.98 941	3	24
37	9.83 931	56	9.84 992	59	0.15 008	9.98 938	2	23
38	9.83 987	56	9.85 051	60	0.14 949	9.98 936	3	22
39	9.84 043	57	9.85 111	59	0.14 889	9.98 933	3	21
40	9.84 100	56	9.85 170	59	0.14 830	9.98 930	3	20
41	9.84 156	56	9.85 229	59	0.14 771	9.98 927	3	19
42	9.84 212	56	9.85 288	59	0.14 712	9.98 924	3	18
43	9.84 268	56	9.85 347	58	0.14 653	9.98 921	2	17
44	9.84 324	56	9.85 405	59	0.14 595	9.98 919	3	16
45	9.84 380	56	9.85 464	59	0.14 536	9.98 916	3	15
46	9.84 436	55	9.85 523	58	0.14 477	9.98 913	3	14
47	9.84 491	56	9.85 581	59	0.14 419	9.98 910	3	13
48	9.84 547	55	9.85 640	58	0.14 360	9.98 907	3	12
49	9.84 602	56	9.85 698	59	0.14 302	9.98 904	3	11
50	9.84 658	55	9.85 757	58	0.14 243	9.98 901	3	10
51	9.84 713	56	9.85 815	58	0.14 185	9.98 898	2	9
52	9.84 769	55	9.85 873	58	0.14 127	9.98 896	3	8
53	9.84 824	55	9.85 931	58	0.14 069	9.98 893	3	7
54	9.84 879	55	9.85 989	58	0.14 011	9.98 890	3	6
55	9.84 934	55	9.86 047	58	0.13 953	9.98 887	3	5
56	9.84 989	55	9.86 105	58	0.13 895	9.98 884	3	4
57	9.85 044	55	9.86 163	58	0.13 837	9.98 881	3	3
58	9.85 099	55	9.86 221	58	0.13 779	9.98 878	3	2
59	9.85 154	55	9.86 279	57	0.13 721	9.98 875	3	1
60	9.85 209	•	9.86 336		0.13 664	9.98 872		0
	L. Cos.	d.	L. Cotg.	c.d.	L. Tang.	L. Sin.	d.	'

P. P.

	63	62	61
1	1,0	1,0	1,0
2	2,1	2,1	2,0
3	3,2	3,1	3,0
4	4,2	4,1	4,1
5	5,2	5,2	5,1
6	6,3	6,2	6,1
7	7,4	7,2	7,1
8	8,4	8,3	8,1
9	9,4	9,3	9,2
10	10,5	10,8	10,2
20	21,0	20,7	20,8
30	31,5	31,0	30,5
40	42,0	41,3	40,7
50	52,5	51,7	50,8

	60	59	58
1	1,0	1,0	1,0
2	2,0	2,0	1,9
3	3,0	3,0	2,9
4	4,0	3,9	3,9
5	5,0	4,9	4,8
6	6,0	5,9	5,8
7	7,0	6,9	6,8
8	8,0	7,9	7,7
9	9,0	8,8	8,7
10	10,0	9,8	9,7
20	20,0	19,7	19,8
30	30,0	29,5	29,0
40	40,0	39,3	38,7
50	50,0	49,2	48,3

	57	56	55
1	1,0	0,9	0,9
2	1,9	1,9	1,8
3	2,8	2,8	2,8
4	3,8	3,7	3,7
5	4,8	4,7	4,6
6	5,7	5,6	5,5
7	6,6	6,5	6,4
8	7,6	7,5	7,3
9	8,6	8,4	8,2
10	9,5	9,3	9,2
20	19,0	18,7	18,3
30	28,5	28,0	27,5
40	38,0	37,3	36,7
50	47,5	46,7	45,8

′	L. Sin.	d.	L. Tang.	c.d.	L. Cotg.	L. Cos.	d.		P. P.		
0	9.85 209	54	9.86 336	58	0.63 664	9.98 872	3	60			
1	9.85 263	55	9.86 394	58	0.63 606	9.98 869	2	59			
2	9.85 318	55	9.86 452	57	0.63 548	9.98 867	3	58			
3	9.85 373	55	9.86 509	57	0.63 491	9.98 864	3	57			
4	9.85 427	54	9.86 566	58	0.63 434	9.98 861	3	56			
5	9.85 481	55	9.86 624	57	0.63 376	9.98 858	3	55	**57**	**56**	**55**
6	9.85 536	54	9.86 681	57	0.63 319	9.98 855	3	54	1 1,0	0,9	0,9
7	9.85 590	54	9.86 738	57	0.63 262	9.98 852	3	53	2 1,9	1,9	1,8
8	9.85 644	54	9.86 795	57	0.63 205	9.98 849	3	52	3 2,8	2,8	2,8
9	9.85 698	54	9.86 852	57	0.63 148	9.98 846	3	51	4 3,8	3,7	3,7
10	9.85 752	54	9.86 909	57	0.63 091	9.98 843	3	50	5 4,8	4,7	4,6
11	9.85 806	54	9.86 966	57	0.63 034	9.98 840	3	49	6 5,7	5,6	5,5
12	9.85 860	54	9.87 023	57	0.62 977	9.98 837	3	48	7 6,6	6,5	6,4
13	9.85 914	54	9.87 080	57	0.62 920	9.98 834	3	47	8 7,6	7,5	7,3
14	9.85 968	54	9.87 137	56	0.62 863	9.98 831	3	46	9 8,6	8,4	8,2
15	9.86 022	53	9.87 193	57	0.62 807	9.98 828	3	45	10 9,5	9,3	9,2
16	9.86 075	54	9.87 250	56	0.62 750	9.98 825	3	44	20 19,0	18,7	18,3
17	9.86 129	53	9.87 306	57	0.62 694	9.98 822	3	43	30 28,5	28,0	27,5
18	9.86 182	54	9.87 363	56	0.62 637	9.98 819	3	42	40 38,0	37,3	36,7
19	9.86 236	53	9.87 419	57	0.62 581	9.98 816	3	41	50 47,5	46,7	45,8
20	9.86 289	53	9.87 476	56	0.62 524	9.98 813	3	40			
21	9.86 342	53	9.87 532	56	0.62 468	9.98 810	3	39			
22	9.86 395	54	9.87 588	56	0.62 412	9.98 807	3	38			
23	9.86 449	53	9.87 644	56	0.62 356	9.98 804	3	37	**54**	**53**	**52**
24	9.86 502	53	9.87 700	56	0.62 300	9.98 801	3	36	1 0,9	0,9	0,9
25	9.86 555	53	9.87 756	56	0.62 244	9.98 798	3	35	2 1,8	1,8	1,7
26	9.86 608	52	9.87 812	56	0.62 188	9.98 795	3	34	3 2,7	2,6	2,6
27	9.86 660	53	9.87 868	56	0.62 132	9.98 792	3	33	4 3,6	3,5	3,5
28	9.86 713	53	9.87 924	56	0.62 076	9.98 789	3	32	5 4,5	4,4	4,3
29	9.86 766	53	9.87 980	55	0.62 020	9.98 786	3	31	6 5,4	5,3	5,2
30	9.86 819	52	9.88 035	56	0.61 965	9.98 783	3	30	7 6,3	6,2	6,1
31	9.86 871	53	9.88 091	56	0.61 909	9.98 780	3	29	8 7,2	7,1	6,9
32	9.86 924	52	9.88 147	55	0.61 853	9.98 777	3	28	9 8,1	8,0	7,8
33	9.86 976	52	9.88 202	55	0.61 798	9.98 774	3	27	10 9,0	8,8	8,7
34	9.87 028	53	9.88 257	56	0.61 743	9.98 771	3	26	20 18,0	17,7	17,3
35	9.87 081	52	9.88 313	55	0.61 687	9.98 768	3	25	30 27,0	26,5	26,0
36	9.87 133	52	9.88 368	55	0.61 632	9.98 765	3	24	40 36,0	35,3	34,7
37	9.87 185	52	9.88 423	56	0.61 577	9.98 762	3	23	50 45,0	44,2	43,3
38	9.87 237	52	9.88 479	55	0.61 521	9.98 759	3	22			
39	9.87 289	52	9.88 534	55	0.61 466	9.98 756	3	21			
40	9.87 341	52	9.88 589	55	0.61 411	9.98 753	3	20			
41	9.87 393	52	9.88 644	55	0.61 356	9.98 750	4	19	**51** **4**	**3**	**2**
42	9.87 445	52	9.88 699	55	0.61 301	9.98 746	3	18	1 0,8 0,1	0,0	0,0
43	9.87 497	52	9.88 754	54	0.61 246	9.98 743	3	17	2 1,7 0,1	0,1	0,1
44	9.87 549	51	9.88 808	55	0.61 192	9.98 740	3	16	3 2,6 0,2	0,2	0,1
45	9.87 600	52	9.88 863	55	0.61 137	9.98 737	3	15	4 3,4 0,3	0,2	0,1
46	9.87 652	51	9.88 918	54	0.61 082	9.98 734	3	14	5 4,2 0,3	0,2	0,2
47	9.87 703	52	9.88 972	55	0.61 028	9.98 731	3	13	6 5,1 0,4	0,3	0,2
48	9.87 755	51	9.89 027	55	0.60 973	9.98 728	3	12	7 6,0 0,5	0,4	0,2
49	9.87 806	52	9.89 082	54	0.60 918	9.98 725	3	11	8 6,8 0,5	0,4	0,3
50	9.87 858	51	9.89 136	54	0.60 864	9.98 722	3	10	9 7,6 0,6	0,4	0,3
51	9.87 909	51	9.89 190	55	0.60 810	9.98 719	4	9	10 8,5 0,7	0,5	0,3
52	9.87 960	51	9.89 245	54	0.60 755	9.98 715	3	8	20 17,0 1,3	1,0	0,7
53	9.88 011	51	9.89 299	54	0.60 701	9.98 712	3	7	30 25,5 2,0	1,5	1,0
54	9.88 062	51	9.89 353	54	0.60 647	9.98 709	3	6	40 34,0 2,7	2,0	1,3
55	9.88 113	51	9.89 407	54	0.60 593	9.98 706	3	5	50 42,5 3,3	2,5	1,7
56	9.88 164	51	9.89 461	54	0.60 539	9.98 703	3	4			
57	9.88 215	51	9.89 515	54	0.60 485	9.98 700	3	3			
58	9.88 266	51	9.89 569	54	0.60 431	9.98 697	3	2			
59	9.88 317	51	9.89 623	54	0.60 377	9.98 694	4	1			
60	9.88 368		9.89 677		0.60 323	9.98 690		0			
	L. Cos.	d.	L. Cotg.	c.d.	L. Tang.	L. Sin.	d.	′	P. P.		

14°

'	L. Sin.	d.	L. Tang.	c.d.	L. Cotg.	L. Cos.	d.	'
0	9.38 368	50	9.89 677	54	0.60 328	9.98 690	3	60
1	9.38 418	51	9.89 731	54	0.60 269	9.98 687	3	59
2	9.38 469	51	9.89 785	53	0.60 215	9.98 684	3	58
3	9.38 519	50	9.89 838	54	0.60 162	9.98 681	3	57
4	9.38 570	50	9.89 892	53	0.60 108	9.98 678	3	56
5	9.38 620	50	9.89 945	54	0.60 055	9.98 675	4	55
6	9.38 670	51	9.89 999	53	0.60 001	9.98 671	3	54
7	9.38 721	50	9.40 052	54	0.59 948	9.98 668	3	53
8	9.38 771	50	9.40 106	53	0.59 894	9.98 665	3	52
9	9.38 821	50	9.40 159	53	0.59 841	9.98 662	3	51
10	9.38 871	50	9.40 212	54	0.59 788	9.98 659	3	50
11	9.38 921	50	9.40 266	53	0.59 734	9.98 656	4	49
12	9.38 971	50	9.40 319	53	0.59 681	9.98 652	3	48
13	9.39 021	50	9.40 372	53	0.59 628	9.98 649	3	47
14	9.39 071	50	9.40 425	53	0.59 575	9.98 646	3	46
15	9.39 121	49	9.40 478	53	0.59 522	9.98 643	3	45
16	9.39 170	50	9.40 531	53	0.59 469	9.98 640	4	44
17	9.39 220	50	9.40 584	52	0.59 416	9.98 636	3	43
18	9.39 270	49	9.40 636	53	0.59 364	9.98 633	3	42
19	9.39 319	50	9.40 689	53	0.59 311	9.98 630	3	41
20	9.39 369	49	9.40 742	53	0.59 258	9.98 627	4	40
21	9.39 418	49	9.40 795	52	0.59 205	9.98 623	3	39
22	9.39 467	50	9.40 847	53	0.59 153	9.98 620	3	38
23	9.39 517	49	9.40 900	52	0.59 100	9.98 617	3	37
24	9.39 566	49	9.40 952	53	0.59 048	9.98 614	4	36
25	9.39 615	49	9.41 005	52	0.58 995	9.98 610	3	35
26	9.39 664	49	9.41 057	52	0.58 943	9.98 607	3	34
27	9.39 713	49	9.41 109	52	0.58 891	9.98 604	3	33
28	9.39 762	49	9.41 161	53	0.58 839	9.98 601	4	32
29	9.39 811	49	9.41 214	52	0.58 786	9.98 597	3	31
30	9.39 860	49	9.41 266	52	0.58 734	9.98 594	3	30
31	9.39 909	49	9.41 318	52	0.58 682	9.98 591	3	29
32	9.39 958	48	9.41 370	52	0.58 630	9.98 588	4	28
33	9.40 006	49	9.41 422	52	0.58 578	9.98 584	3	27
34	9.40 055	48	9.41 474	52	0.58 526	9.98 581	3	26
35	9.40 103	49	9.41 526	52	0.58 474	9.98 578	4	25
36	9.40 152	48	9.41 578	51	0.58 422	9.98 574	3	24
37	9.40 200	49	9.41 629	52	0.58 371	9.98 571	3	23
38	9.40 249	48	9.41 681	52	0.58 319	9.98 568	3	22
39	9.40 297	49	9.41 733	51	0.58 267	9.98 565	4	21
40	9.40 346	48	9.41 784	52	0.58 216	9.98 561	3	20
41	9.40 394	48	9.41 836	51	0.58 164	9.98 558	3	19
42	9.40 442	48	9.41 887	52	0.58 113	9.98 555	4	18
43	9.40 490	48	9.41 939	51	0.58 061	9.98 551	3	17
44	9.40 538	48	9.41 990	51	0.58 010	9.98 548	3	16
45	9.40 586	48	9.42 041	52	0.57 959	9.98 545	4	15
46	9.40 634	48	9.42 093	51	0.57 907	9.98 541	3	14
47	9.40 682	48	9.42 144	51	0.57 856	9.98 538	3	13
48	9.40 730	48	9.42 195	51	0.57 805	9.98 535	3	12
49	9.40 778	47	9.42 246	51	0.57 754	9.98 531	4	11
50	9.40 825	48	9.42 297	51	0.57 703	9.98 528	3	10
51	9.40 873	48	9.42 348	51	0.57 652	9.98 525	4	9
52	9.40 921	47	9.42 399	51	0.57 601	9.98 521	3	8
53	9.40 968	48	9.42 450	51	0.57 550	9.98 518	3	7
54	9.41 016	47	9.42 501	51	0.57 499	9.98 515	3	6
55	9.41 063	48	9.42 552	51	0.57 448	9.98 511	4	5
56	9.41 111	47	9.42 603	50	0.57 397	9.98 508	3	4
57	9.41 158	47	9.42 653	51	0.57 347	9.98 505	3	3
58	9.41 205	47	9.42 704	51	0.57 296	9.98 501	4	2
59	9.41 252	48	9.42 755	50	0.57 245	9.98 498	3	1
60	9.41 300		9.42 805		0.57 195	9.98 494		0

L. Cos.	d.	L. Cotg.	c.d.	L. Tang.	L. Sin.	d.	'

P. P.

	54	53	52
1	0,9	0,9	0,9
2	1,8	1,8	1,7
3	2,7	2,6	2,6
4	3,6	3,5	3,5
5	4,5	4,4	4,3
6	5,4	5,3	5,2
7	6,3	6,2	6,1
8	7,2	7,1	6,9
9	8,1	8,0	7,8
10	9,0	8,8	8,7
20	18,0	17,7	17,8
30	27,0	26,5	26,0
40	36,0	35,3	34,7
50	45,0	44,2	43,3

	51	50	49
1	0,8	0,8	0,8
2	1,7	1,7	1,6
3	2,6	2,5	2,4
4	3,4	3,3	3,3
5	4,2	4,2	4,1
6	5,1	5,0	4,9
7	6,0	5,8	5,7
8	6,8	6,7	6,5
9	7,6	7,5	7,4
10	8,5	8,3	8,2
20	17,0	16,7	16,3
30	25,5	25,0	24,5
40	34,0	33,3	32,7
50	42,5	41,7	40,8

	48	47	4	3
1	0,8	0,8	0,1	0,0
2	1,6	1,6	0,1	0,1
3	2,4	2,4	0,2	0,1
4	3,2	3,1	0,3	0,2
5	4,0	3,9	0,3	0,2
6	4,8	4,7	0,4	0,3
7	5,6	5,5	0,5	0,4
8	6,4	6,3	0,5	0,4
9	7,2	7,0	0,6	0,4
10	8,0	7,8	0,7	0,5
20	16,0	15,7	1,3	1,0
30	24,0	23,5	2,0	1,5
40	32,0	31,3	2,7	2,0
50	40,0	39,2	3,3	2,5

75°

15°

′	L. Sin.	d.	L. Tang.	c.d.	L. Cotg.	L. Cos.	d.		P. P.		
0	9.41 800	47	9.42 805	51	0.57 195	9.98 494	3	60			
1	9.41 847	47	9.42 856	50	0.57 144	9.98 491	3	59			
2	9.41 894	47	9.42 906	51	0.57 094	9.98 488	4	58			
3	9.41 441	47	9.42 957	50	0.57 043	9.98 484	3	57			
4	9.41 488	47	9.43 007	50	0.56 993	9.98 481	4	56		51	50	49
5	9.41 535	47	9.43 057	51	0.56 943	9.98 477	3	55	1	0,8	0,8	0,8
6	9.41 582	46	9.43 108	50	0.56 892	9.98 474	3	54	2	1,7	1,7	
7	9.41 628	47	9.43 158	50	0.56 842	9.98 471	3	53	3	2,6	2,5	
8	9.41 675	47	9.43 208	50	0.56 792	9.98 467	3	52	4	3,4	3,3	
9	9.41 722	46	9.43 258	50	0.56 742	9.98 464	4	51				
10	9.41 768	47	9.43 308	50	0.56 692	9.98 460	3	50	5	4,2	4,2	4,1
11	9.41 815	46	9.43 358	50	0.56 642	9.98 457	4	49	6	5,1	5,0	4,9
12	9.41 861	47	9.43 408	50	0.56 592	9.98 453	3	48	7	6,0	5,8	5,7
13	9.41 908	46	9.43 458	50	0.56 542	9.98 450	4	47	8	6,8	6,7	6,5
14	9.41 954	47	9.43 508	50	0.56 492	9.98 447	4	46	9	7,6	7,5	7,4
15	9.42 001	46	9.43 558	49	0.56 442	9.98 443	3	45	10	8,5	8,3	8,2
16	9.42 047	46	9.43 607	50	0.56 393	9.98 440	4	44	20	17,0	26,6	16,3
17	9.42 098	47	9.43 657	50	0.56 343	9.98 436	3	43	30	25,5	5	24
18	9.42 140	46	9.43 707	50	0.56 293	9.98 433	4	42	40	34,0	33	32
19	9.42 186	46	9.43 756	49	0.56 244	9.98 429	3	41	50	42,5	41,6	40
20	9.42 232	46	9.43 806	50	0.56 194	9.98 426	4	40				
21	9.42 278	46	9.43 855	49	0.56 145	9.98 422	4	39				
22	9.42 324	46	9.43 905	50	0.56 095	9.98 419	4	38				
23	9.42 370	46	9.43 954	50	0.56 046	9.98 415	3	37		48	47	46
24	9.42 416	45	9.44 004	49	0.55 996	9.98 412	3	36				
25	9.42 461	46	9.44 058	49	0.55 947	9.98 409	4	35	1	0,8	0,8	0,8
26	9.42 507	46	9.44 102	49	0.55 898	9.98 405	3	34	3	1,6	2,4	2,5
27	9.42 553	46	9.44 151	50	0.55 849	9.98 402	4	33		4		3,8
28	9.42 599	45	9.44 201	49	0.55 799	9.98 398	3	32		2		1
29	9.42 644	46	9.44 250	49	0.55 750	9.98 395	4	31	5	4,0	3,9	3,8
30	9.42 690	45	9.44 299	49	0.55 701	9.98 391	3	30		8	4,7	
31	9.42 735	46	9.44 348	49	0.55 652	9.98 388	4	29		6	5,5	
32	9.42 781	45	9.44 397	49	0.55 603	9.98 384	3	28		4	6,3	
33	9.42 826	46	9.44 446	49	0.55 554	9.98 381	4	27		2	7,0	
34	9.42 872	45	9.44 495	49	0.55 505	9.98 377	4	26				
35	9.42 917	45	9.44 544	48	0.55 456	9.98 373	3	25	10	8,0	7,8	7,7
36	9.42 962	46	9.44 592	49	0.55 408	9.98 370	4	24	20	16,0	15,7	15,3
37	9.43 008	45	9.44 641	49	0.55 359	9.98 366	3	23	30	24,0	23,5	23,0
38	9.43 053	45	9.44 690	48	0.55 310	9.98 363	4	22	40	32,0	31,3	30,7
39	9.43 098	45	9.44 738	49	0.55 262	9.98 359	3	21	50	40,	39,2	38,0
40	9.43 143	45	9.44 787	49	0.55 213	9.98 356	4	20				
41	9.43 188	45	9.44 836	48	0.55 164	9.98 352	3	19				
42	9.43 233	45	9.44 884	49	0.55 116	9.98 349	4	18				
43	9.43 278	45	9.44 933	48	0.55 067	9.98 345	3	17		45	44	4 3
44	9.43 323	44	9.44 981	48	0.55 019	9.98 342	4	16				
45	9.43 367	45	9.45 029	49	0.54 971	9.98 338	4	15	1	0,8	0,7	0,1 0,0
46	9.43 412	45	9.45 078	48	0.54 922	9.98 334	3	14	2	1,5	2,5	0,1 0,2
47	9.43 457	45	9.45 126	48	0.54 874	9.98 331	4	13	3	2	2,0 2	
48	9.43 502	44	9.45 174	48	0.54 826	9.98 327	3	12	4	3	9 0,8	
49	9.43 546	45	9.45 222	49	0.54 778	9.98 324	4	11	5	3,8	3,7 0	2,0 9
50	9.43 591	44	9.45 271	48	0.54 729	9.98 320	3	10		4,5	4,4 0,5	3 0,4
51	9.43 635	45	9.45 319	48	0.54 681	9.98 317	4	9		5	,1 0,	
52	9.43 680	44	9.45 367	48	0.54 633	9.98 313	4	8		6	,9 0,	
53	9.43 724	45	9.45 415	48	0.54 585	9.98 309	3	7		6	,6 0,	
54	9.43 769	44	9.45 463	48	0.54 537	9.98 306	4	6	10	7,5	7,8 0,7	0 5
55	9.43 813	44	9.45 511	48	0.54 489	9.98 302	3	5	20	15,0	14,7 1,8	2,0
56	9.43 857	44	9.45 559	47	0.54 441	9.98 299	4	4	30	32,0	22 0 2,0	
57	9.43 901	45	9.45 606	48	0.54 394	9.98 295	4	3	40	0	29 8 2,7	
58	9.43 946	44	9.45 654	48	0.54 346	9.98 291	4	2	50	7	36 7 3,8	
59	9.43 990	44	9.45 702	48	0.54 298	9.98 288	4	1				
60	9.44 034		9.45 750		0.54 250	9.98 284		0				

	L. Cos.	d.	L. Cotg.	c.d.	L. Tang.	L. Sin.	d.	′		P. P.	

16°

′	L. Sin.	d.	L. Tang.	c.d.	L. Cotg.	L. Cos.	d.		P. P.		
0	9.44 084		9.45 750		0.54 250	9.98 284		60			
1	9.44 078	44	9.45 797	47	0.54 203	9.98 281	8	59			
2	9.44 122	44	9.45 845	48	0.54 155	9.98 277	4	58			
3	9.44 166	44	9.45 892	47	0.54 108	9.98 273	4	57			
4	9.44 210	44	9.45 940	48	0.54 060	9.98 270	3	56	48	47	46
		43		47			4				
5	9.44 253	44	9.45 987	48	0.54 013	9.98 266	4	55	1	0,8 0,8 0,8
6	9.44 297	44	9.46 085	47	0.53 965	9.98 262	3	54	2	1,6 1,6 1,5
7	9.44 341	44	9.46 082	48	0.53 918	9.98 259	4	53	3	2,4 2,4 2,3
8	9.44 385	43	9.46 130	47	0.53 870	9.98 255	4	52	4	3,2 3,1 3,1
9	9.44 428	44	9.46 177	47	0.53 823	9.98 251	3	51		
10	9.44 472	44	9.46 224	47	0.53 776	9.98 248	4	50	5	4,0 3,9 3,8
11	9.44 516	43	9.46 271	48	0.53 729	9.98 244	4	49	6	4,8 4,7 4,6
12	9.44 559	43	9.46 319	47	0.53 681	9.98 240	3	48	7	5,6 5,5 5,4
13	9.44 602	44	9.46 366	47	0.53 634	9.98 237	4	47	8	6,4 6,3 6,1
14	9.44 646	43	9.46 413	47	0.53 587	9.98 233	4	46	9	7,2 7,0 6,9
15	9.44 689	44	9.46 460	47	0.53 540	9.98 229	3	45	10	8,0 7,8 7,7
16	9.44 733	43	9.46 507	47	0.53 493	9.98 226	4	44	20	16,0 15,7 15,8
17	9.44 776	43	9.46 554	47	0.53 446	9.98 222	4	43	30	24,0 23,5 23,0
18	9.44 819	43	9.46 601	47	0.53 399	9.98 218	3	42	40	32,0 31,3 30,7
19	9.44 862	43	9.46 648	46	0.53 352	9.98 215	4	41	50	40,0 39,2 38,3
20	9.44 905	43	9.46 694	47	0.53 306	9.98 211	4	40		
21	9.44 948	44	9.46 741	47	0.53 259	9.98 207	3	39		
22	9.44 992	43	9.46 788	47	0.53 212	9.98 204	4	38		
23	9.45 035	42	9.46 835	46	0.53 165	9.98 200	4	37	45	44	43
24	9.45 077	43	9.46 881	47	0.53 119	9.98 196	4	36		
25	9.45 120	43	9.46 928	47	0.53 072	9.98 192	3	35	1	0,8 0,7 0,7
26	9.45 163	43	9.46 975	46	0.53 025	9.98 189	4	34	2	1,5 1,5 1,4
27	9.45 206	43	9.47 021	47	0.52 979	9.98 185	4	33	3	2,2 2,2 2,2
28	9.45 249	43	9.47 068	46	0.52 932	9.98 181	4	32	4	3,0 2,9 2,9
29	9.45 292	42	9.47 114	46	0.52 886	9.98 177	3	31		
30	9.45 334	43	9.47 160	47	0.52 840	9.98 174	4	30	5	3,8 3,7 3,6
31	9.45 377	42	9.47 207	46	0.52 793	9.98 170	4	29	6	4,5 4,4 4,3
32	9.45 419	43	9.47 253	46	0.52 747	9.98 166	4	28	7	5,2 5,1 5,0
33	9.45 462	42	9.47 299	47	0.52 701	9.98 162	3	27	8	6,0 5,9 5,7
34	9.45 504	43	9.47 346	46	0.52 654	9.98 159	4	26	9	6,8 6,6 6,4
35	9.45 547	42	9.47 392	46	0.52 608	9.98 155	4	25	10	7,5 7,3 7,2
36	9.45 589	43	9.47 438	46	0.52 562	9.98 151	4	24	20	15,0 14,7 14,3
37	9.45 632	42	9.47 484	46	0.52 516	9.98 147	3	23	30	22,5 22,0 21,5
38	9.45 674	42	9.47 530	46	0.52 470	9.98 144	4	22	40	30,0 29,3 28,7
39	9.45 716	42	9.47 576	46	0.52 424	9.98 140	4	21	50	37,5 36,7 35,8
40	9.45 758	43	9.47 622	46	0.52 378	9.98 136	4	20		
41	9.45 801	42	9.47 668	46	0.52 332	9.98 132	3	19		
42	9.45 843	42	9.47 714	46	0.52 286	9.98 129	4	18		
43	9.45 885	42	9.47 760	46	0.52 240	9.98 125	4	17	42	41	4	3
44	9.45 927	42	9.47 806	46	0.52 194	9.98 121	4	16		
45	9.45 969	42	9.47 852	45	0.52 148	9.98 117	3	15	1	0,7 0,7 0,1 0,0
46	9.46 011	42	9.47 897	46	0.52 103	9.98 113	4	14	2	1,4 1,4 0,1 0,1
47	9.46 053	42	9.47 943	46	0.52 057	9.98 110	3	13	3	2,1 2,0 0,2 0,2
48	9.46 095	41	9.47 989	46	0.52 011	9.98 106	4	12	4	2,8 2,7 0,3 0,2
49	9.46 136	42	9.48 035	45	0.51 965	9.98 102	4	11		
50	9.46 178	42	9.48 080	46	0.51 920	9.98 098	4	10	5	3,5 3,4 0,3 0,2
51	9.46 220	42	9.48 126	45	0.51 874	9.98 094	4	9	6	4,2 4,1 0,4 0,3
52	9.46 262	41	9.48 171	46	0.51 829	9.98 090	3	8	7	4,9 4,8 0,5 0,4
53	9.46 303	42	9.48 217	45	0.51 783	9.98 087	4	7	8	5,6 5,5 0,5 0,4
54	9.46 345	41	9.48 262	45	0.51 738	9.98 083	4	6	9	6,3 6,2 0,6 0,4
55	9.46 386	42	9.48 307	46	0.51 693	9.98 079	4	5	10	7,0 6,8 0,7 0,5
56	9.46 428	41	9.48 353	45	0.51 647	9.98 075	4	4	20	14,0 13,7 1,8 1,0
57	9.46 469	42	9.48 398	45	0.51 602	9.98 071	3	3	30	21,0 20,5 2,0 1,5
58	9.46 511	41	9.48 443	46	0.51 557	9.98 067	4	2	40	28,0 27,3 2,7 2,0
59	9.46 552	42	9.48 489	45	0.51 511	9.98 063	4	1	50	35,0 34,2 3,3 2,5
60	9.46 594		9.48 534		0.51 466	9.98 060		0		
	L. Cos.	d.	L. Cotg.	c.d.	L. Tang.	L. Sin.	d.	′		P. P.	

73°

17°

′	L. Sin.	d.	L. Tang.	c.d.	L. Cotg.	L. Cos.	d.		P. P.		
0	9.46 594	41	9.48 584	45	0.51 466	9.98 060	4	60			
1	9.46 635	41	9.48 579	45	0.51 421	9.98 056	4	59			
2	9.46 676	41	9.48 624	45	0.51 376	9.98 052	4	58			
3	9.46 717	41	9.48 669	45	0.51 331	9.98 048	4	57			
4	9.46 758	42	9.48 714	45	0.51 286	9.98 044	4	56		**45**	**44**	**43**
5	9.46 800	41	9.48 759	45	0.51 241	9.98 040	4	55	1	0,8	0,7	0,7
6	9.46 841	41	9.48 804	45	0.51 196	9.98 036	4	54	2	1,5	1,5	1,4
7	9.46 882	41	9.48 849	45	0.51 151	9.98 032	3	53	3	2,2	2,2	2,2
8	9.46 923	41	9.48 894	45	0.51 106	9.98 029	4	52	4	3,0	2,9	2,9
9	9.46 964	41	9.48 939	45	0.51 061	9.98 025	4	51				
10	9.47 005	40	9.48 984	45	0.51 016	9.98 021	4	50	5	3,8	3,7	3,6
11	9.47 045	41	9.49 029	44	0.50 971	9.98 017	4	49	6	4,5	4,4	4,3
12	9.47 086	41	9.49 073	45	0.50 927	9.98 013	4	48	7	5,2	5,1	5,0
13	9.47 127	41	9.49 118	45	0.50 882	9.98 009	4	47	8	6,0	5,9	5,7
14	9.47 168	41	9.49 163	44	0.50 837	9.98 005	4	46	9	6,8	6,6	6,4
15	9.47 209	40	9.49 207	45	0.50 793	9.98 001	4	45	10	7,5	7,3	7,2
16	9.47 249	41	9.49 252	44	0.50 748	9.97 997	4	44	20	15,0	14,7	14,3
17	9.47 290	40	9.49 296	45	0.50 704	9.97 993	4	43	30	22,5	22,0	21,5
18	9.47 330	41	9.49 341	44	0.50 659	9.97 989	3	42	40	30,0	29,3	28,7
19	9.47 371	40	9.49 385	45	0.50 615	9.97 986	4	41	50	37,5	36,7	35,8
20	9.47 411	41	9.49 430	44	0.50 570	9.97 982	4	40				
21	9.47 452	40	9.49 474	45	0.50 526	9.97 978	4	39				
22	9.47 492	41	9.49 519	44	0.50 481	9.97 974	4	38				
23	9.47 533	40	9.49 563	44	0.50 437	9.97 970	4	37		**42**	**41**	**40**
24	9.47 573	40	9.49 607	45	0.50 393	9.97 966	4	36				
25	9.47 613	41	9.49 652	44	0.50 348	9.97 962	4	35	1	0,7	0,7	0,7
26	9.47 654	40	9.49 696	44	0.50 304	9.97 958	4	34	2	1,4	1,4	1,3
27	9.47 694	40	9.49 740	44	0.50 260	9.97 954	4	33	3	2,1	2,0	2,0
28	9.47 734	40	9.49 784	44	0.50 216	9.97 950	4	32	4	2,8	2,7	2,7
29	9.47 774	40	9.49 828	44	0.50 172	9.97 946	4	31				
30	9.47 814	40	9.49 872	44	0.50 128	9.97 942	4	30	5	3,5	3,4	3,3
31	9.47 854	40	9.49 916	44	0.50 084	9.97 938	4	29	6	4,2	4,1	4,0
32	9.47 894	40	9.49 960	44	0.50 040	9.97 934	4	28	7	4,9	4,8	4,7
33	9.47 934	40	9.50 004	44	0.49 996	9.97 930	4	27	8	5,6	5,5	5,3
34	9.47 974	40	9.50 048	44	0.49 952	9.97 926	4	26	9	6,3	6,2	6,0
35	9.48 014	40	9.50 092	44	0.49 908	9.97 922	4	25	10	7,0	6,8	6,7
36	9.48 054	40	9.50 136	44	0.49 864	9.97 918	4	24	20	14,0	13,7	13,3
37	9.48 094	39	9.50 180	43	0.49 820	9.97 914	4	23	30	21,0	20,5	20,0
38	9.48 133	40	9.50 223	44	0.49 777	9.97 910	4	22	40	28,0	27,3	26,7
39	9.48 173	40	9.50 267	44	0.49 733	9.97 906	4	21	50	35,0	34,2	33,3
40	9.48 213	39	9.50 311	44	0.49 689	9.97 902	4	20				
41	9.48 252	40	9.50 355	43	0.49 645	9.97 898	4	19				
42	9.48 292	40	9.50 398	44	0.49 602	9.97 894	4	18				
43	9.48 332	40	9.50 442	43	0.49 558	9.97 890	4	17		**39**	**5**	**4**	**3**
44	9.48 371	40	9.50 485	44	0.49 515	9.97 886	4	16	1	0,6 0,1	0,1	0,0
45	9.48 411	39	9.50 529	43	0.49 471	9.97 882	4	15	2	1,3 0,2	0,1	0,1
46	9.48 450	40	9.50 572	44	0.49 428	9.97 878	4	14	3	2,0 0,2	0,2	0,2
47	9.48 490	39	9.50 616	43	0.49 384	9.97 874	4	13	4	2,6 0,3	0,3	0,2
48	9.48 529	39	9.50 659	44	0.49 341	9.97 870	4	12				
49	9.48 568	39	9.50 703	43	0.49 297	9.97 866	5	11	5	3,2 0,4	0,3	0,2
									6	3,9 0,5	0,4	0,3
50	9.48 607	40	9.50 746	43	0.49 254	9.97 861	4	10	7	4,6 0,6	0,5	0,4
51	9.48 647	39	9.50 789	44	0.49 211	9.97 857	4	9	8	5,2 0,7	0,5	0,4
52	9.48 686	39	9.50 833	43	0.49 167	9.97 853	4	8	9	5,8 0,8	0,6	0,4
53	9.48 725	39	9.50 876	43	0.49 124	9.97 849	4	7				
54	9.48 764	39	9.50 919	43	0.49 081	9.97 845	4	6	10	6,5 0,8	0,7	0,5
55	9.48 803	39	9.50 962	43	0.49 038	9.97 841	4	5	20	13,0 1,7	1,3	1,0
56	9.48 842	39	9.51 005	43	0.48 995	9.97 837	4	4	30	19,5 2,5	2,0	1,5
57	9.48 881	39	9.51 048	44	0.48 952	9.97 833	4	3	40	26,0 3,3	2,7	2,0
58	9.48 920	39	9.51 092	43	0.48 908	9.97 829	4	2	50	32,5 4,2	3,3	2,5
59	9.48 959	39	9.51 135	43	0.48 865	9.97 825	4	1				
60	9.48 998		9.51 178		0.48 822	9.97 821		0				
	L. Cos.	d.	L. Cotg.	c.d.	L. Tang.	L. Sin.	d.	′		P. P.		

72°

18°

'	L. Sin.	d.	L. Tang.	c.d.	L. Cotg.	L. Cos.	d.		P. P.		
0	9.48 998		9.51 178		0.48 822	9.97 821	4	60			
1	9.49 087	89	9.51 221	43	0.48 779	9.97 817	5	59			
2	9.49 076	89	9.51 264	43	0.48 736	9.97 812	4	58			
3	9.49 115	89	9.51 306	42	0.48 694	9.97 808	4	57			
4	9.49 153	38	9.51 349	43	0.48 651	9.97 804	4	56		**43**	**42**	**41**
5	9.49 192	89	9.51 392	43	0.48 608	9.97 800	4	55	1	0,7	0,7	0,7
6	9.49 231	89	9.51 435	43	0.48 565	9.97 796	4	54	2	1,4	1,4	1,4
7	9.49 269	88	9.51 478	43	0.48 522	9.97 792	4	53	8	2,2	2,1	2,0
8	9.49 308	89	9.51 520	43	0.48 480	9.97 788	4	52	4	2,9	2,8	2,7
9	9.49 347	88	9.51 563	43	0.48 437	9.97 784	5	51			
10	9.49 385	89	9.51 606	42	0.48 394	9.97 779	4	50	5	3,6	3,5	3,4
11	9.49 424	38	9.51 648	43	0.48 352	9.97 775	4	49	6	4,3	4,2	4,1
12	9.49 462	38	9.51 691	43	0.48 309	9.97 771	4	48	7	5,0	4,9	4,8
13	9.49 500	89	9.51 734	42	0.48 266	9.97 767	4	47	8	5,7	5,6	5,5
14	9.49 539	38	9.51 776	43	0.48 224	9.97 763	4	46	9	6,4	6,8	6,2
15	9.49 577	38	9.51 819	42	0.48 181	9.97 759	5	45	10	7,2	7,0	6,8
16	9.49 615	89	9.51 861	42	0.48 139	9.97 754	4	44	20	14,3	14,0	18,7
17	9.49 654	88	9.51 903	43	0.48 097	9.97 750	4	43	30	21,5	21,0	20,5
18	9.49 692	38	9.51 946	42	0.48 054	9.97 746	4	42	40	28,7	28,0	27,8
19	9.49 730	38	9.51 988	43	0.48 012	9.97 742	4	41	50	35,8	35,0	34,2
20	9.49 768	38	9.52 031	42	0.47 969	9.97 738	4	40			
21	9.49 806	38	9.52 073	42	0.47 927	9.97 734	5	39			
22	9.49 844	38	9.52 115	42	0.47 885	9.97 729	4	38			
23	9.49 882	38	9.52 157	43	0.47 843	9.97 725	4	37		**39**	**38**	**37**
24	9.49 920	38	9.52 200	42	0.47 800	9.97 721	4	36			
25	9.49 958	38	9.52 242	42	0.47 758	9.97 717	4	35	1	0,6	0,6	0,6
26	9.49 996	38	9.52 284	42	0.47 716	9.97 713	5	34	2	1,8	1,8	1,2
27	9.50 034	38	9.52 326	42	0.47 674	9.97 708	4	33	8	2,0	1,9	1,8
28	9.50 072	38	9.52 368	42	0.47 632	9.97 704	4	32	4	2,6	2,5	2,5
29	9.50 110	38	9.52 410	42	0.47 590	9.97 700	4	31	5	3,2	3,2	3,1
30	9.50 148	37	9.52 452	42	0.47 548	9.97 696	5	30	6	3,9	3,8	3,7
31	9.50 185	38	9.52 494	42	0.47 506	9.97 691	4	29	7	4,6	4,4	4,3
32	9.50 223	38	9.52 536	42	0.47 464	9.97 687	4	28	8	5,2	5,1	4,9
33	9.50 261	37	9.52 578	42	0.47 422	9.97 683	4	27	9	5,8	5,7	5,6
34	9.50 298	38	9.52 620	41	0.47 380	9.97 679	5	26			
35	9.50 336	38	9.52 661	42	0.47 339	9.97 674	4	25	10	6,5	6,3	6,2
36	9.50 374	37	9.52 703	42	0.47 297	9.97 670	4	24	20	13,0	12,7	12,3
37	9.50 411	38	9.52 745	42	0.47 255	9.97 666	4	23	30	19,5	19,0	18,5
38	9.50 449	37	9.52 787	42	0.47 213	9.97 662	5	22	40	26,0	25,3	24,7
39	9.50 486	37	9.52 829	41	0.47 171	9.97 657	4	21	50	32,5	31,7	30,8
40	9.50 523	38	9.52 870	42	0.47 130	9.97 653	4	20			
41	9.50 561	37	9.52 912	41	0.47 088	9.97 649	4	19			
42	9.50 598	37	9.52 953	42	0.47 047	9.97 645	5	18			
43	9.50 635	38	9.52 995	42	0.47 005	9.97 640	4	17		**36**	**5**	**4**
44	9.50 673	37	9.53 037	41	0.46 963	9.97 636	4	16	1	0,6	0,1	0,1
45	9.50 710	37	9.53 078	42	0.46 922	9.97 632	4	15	2	1,2	0,2	0,1
46	9.50 747	37	9.53 120	41	0.46 880	9.97 628	5	14	8	1,8	0,2	0,2
47	9.50 784	37	9.53 161	41	0.46 839	9.97 623	4	13	4	2,4	0,3	0,3
48	9.50 821	37	9.53 202	42	0.46 798	9.97 619	4	12			
49	9.50 858	38	9.53 244	41	0.46 756	9.97 615	5	11	5	3,0	0,4	0,3
50	9.50 896	37	9.53 285	42	0.46 715	9.97 610	4	10	6	3,6	0,5	0,4
51	9.50 933	37	9.53 327	41	0.46 673	9.97 606	4	9	7	4,2	0,6	0,5
52	9.50 970	37	9.53 368	41	0.46 632	9.97 602	5	8	8	4,8	0,7	0,5
53	9.51 007	36	9.53 409	41	0.46 591	9.97 597	4	7	9	5,4	0,8	0,6
54	9.51 043	37	9.53 450	42	0.46 550	9.97 593	4	6			
55	9.51 080	37	9.53 492	41	0.46 508	9.97 589	5	5	10	6,0	0,8	0,7
56	9.51 117	37	9.53 533	41	0.46 467	9.97 584	4	4	20	12,0	1,7	1,3
57	9.51 154	37	9.53 574	41	0.46 426	9.97 580	4	3	30	18,0	2,5	2,0
58	9.51 191	36	9.53 615	41	0.46 385	9.97 576	5	2	40	24,0	3,3	2,3
59	9.51 227	37	9.53 656	41	0.46 344	9.97 571	4	1	50	30,0	4,2	3,3
60	9.51 264		9.53 697		0.46 303	9.97 567		0			
	L. Cos.	d.	L. Cotg.	c.d.	L. Tang.	L. Sin.	d.	'		P. P.		

71°

19°

′	L. Sin.	d.	L. Tang.	c.d.	L. Cotg.	L Cos.	d.	′
0	9.51 264	37	9.58 697	41	0.46 303	9.97 567	4	60
1	9.51 301	37	9.58 738	41	0.46 262	9.97 563	5	59
2	9.51 338	36	9.58 779	41	0.46 221	9.97 558	4	58
3	9.51 374	37	9.58 820	41	0.46 180	9.97 554	4	57
4	9.51 411	36	9.58 861	41	0.46 139	9.97 550	5	56
5	9.51 447	37	9.58 902	41	0.46 098	9.97 545	4	55
6	9.51 484	36	9.58 943	41	0.46 057	9.97 541	5	54
7	9.51 520	37	9.58 984	41	0.46 016	9.97 536	5	53
8	9.51 557	36	9.54 025	40	0.45 975	9.97 532	4	52
9	9.51 593	36	9.54 065	41	0.45 935	9.97 528	5	51
10	9.51 629	37	9.54 106	41	0.45 894	9.97 523		50
11	9.51 666	36	9.54 147	40	0.45 853	9.97 519	4	49
12	9.51 702	36	9.54 187	41	0.45 813	9.97 515	5	48
13	9.51 738	36	9.54 228	41	0.45 772	9.97 510	4	47
14	9.51 774	37	9.54 269	40	0.45 731	9.97 506	5	46
15	9.51 811	36	9.54 309	41	0.45 691	9.97 501	4	45
16	9.51 847	36	9.54 350	40	0.45 650	9.97 497	5	44
17	9.51 883	36	9.54 390	41	0.45 610	9.97 492	4	43
18	9.51 919	36	9.54 431	40	0.45 569	9.97 488	4	42
19	9.51 955	36	9.54 471	41	0.45 529	9.97 484	5	41
20	9.51 991	36	9.54 512	40	0.45 488	9.97 479	4	40
21	9.52 027	36	9.54 552	41	0.45 448	9.97 475	5	39
22	9.52 063	36	9.54 593	40	0.45 407	9.97 470	4	38
23	9.52 099	36	9.54 633	40	0.45 367	9.97 466	5	37
24	9.52 135	36	9.54 673	41	0.45 327	9.97 461	4	36
25	9.52 171	36	9.54 714	40	0.45 286	9.97 457	4	35
26	9.52 207	35	9.54 754	40	0.45 246	9.97 453	5	34
27	9.52 242	36	9.54 794	41	0.45 206	9.97 448	4	33
28	9.52 278	36	9.54 835	40	0.45 165	9.97 444	5	32
29	9.52 314	36	9.54 875	40	0.45 125	9.97 439	4	31
30	9.52 350	35	9.54 915	40	0.45 085	9.97 435	5	30
31	9.52 385	36	9.54 955	40	0.45 045	9.97 430	4	29
32	9.52 421	35	9.54 995	40	0.45 005	9.97 426	4	28
33	9.52 456	36	9.55 035	40	0.44 965	9.97 421	5	27
34	9.52 492	35	9.55 075	40	0.44 925	9.97 417	4	26
35	9.52 527	36	9.55 115	40	0.44 885	9.97 412	4	25
36	9.52 563	35	9.55 155	40	0.44 845	9.97 408	5	24
37	9.52 598	36	9.55 195	40	0.44 805	9.97 403	4	23
38	9.52 634	35	9.55 235	40	0.44 765	9.97 399	5	22
39	9.52 669	36	9.55 275	40	0.44 725	9.97 394	4	21
40	9.52 705	35	9.55 315	40	0.44 685	9.97 390	5	20
41	9.52 740	35	9.55 355	40	0.44 645	9.97 385	4	19
42	9.52 775	36	9.55 395	39	0.44 605	9.97 381	5	18
43	9.52 811	35	9.55 434	40	0.44 566	9.97 376	4	17
44	9.52 846	35	9.55 474	40	0.44 526	9.97 372	5	16
45	9.52 881	35	9.55 514	40	0.44 486	9.97 367	4	15
46	9.52 916	35	9.55 554	39	0.44 446	9.97 363	5	14
47	9.52 951	35	9.55 593	40	0.44 407	9.97 358	5	13
48	9.52 986	35	9.55 633	40	0.44 367	9.97 353	5	12
49	9.53 021	35	9.55 673	39	0.44 327	9.97 349	4	11
50	9.53 056	36	9.55 712	40	0.44 288	9.97 344	4	10
51	9.53 092	34	9.55 752	39	0.44 248	9.97 340	5	9
52	9.53 126	35	9.55 791	40	0.44 209	9.97 335	4	8
53	9.53 161	35	9.55 831	39	0.44 169	9.97 331	5	7
54	9.53 196	35	9.55 870	40	0.44 130	9.97 326	4	6
55	9.53 231	35	9.55 910	39	0.44 090	9.97 322	5	5
56	9.53 266	35	9.55 949	40	0.44 051	9.97 317	5	4
57	9.53 301	35	9.55 989	39	0.44 011	9.97 312	4	3
58	9.53 336	34	9.56 028	39	0.43 972	9.97 308	5	2
59	9.53 370	35	9.56 067	40	0.43 933	9.97 303	4	1
60	9.53 405		9.56 107		0.43 893	9.97 299		0
	L. Cos.	d.	L. Cotg.	c.d.	L. Tang.	L. Sin.	d.	′

P. P.

	41	40	39
1	0,7	0,7	0,6
2	1,4	1,3	1,3
3	2,0	2,0	2,0
4	2,7	2,7	2,6
5	3,4	3,3	3,2
6	4,1	4,0	3,9
7	4,8	4,7	4,6
8	5,5	5,3	5,2
9	6,2	6,0	5,8
10	6,8	6,7	6,5
20	13,7	13,3	13,0
30	20,5	20,0	19,5
40	27,3	26,7	26,0
50	34,2	33,3	32,5

	37	36	35
1	0,6	0,6	0,6
2	1,2	1,2	1,2
3	1,8	1,8	1,8
4	2,5	2,4	2,3
5	3,1	3,0	2,9
6	3,7	3,6	3,5
7	4,3	4,2	4,1
8	4,9	4,8	4,7
9	5,6	5,4	5,2
10	6,2	6,0	5,8
20	12,3	12,0	11,7
30	18,5	18,0	17,5
40	24,7	24,0	23,3
50	30,8	30,0	29,2

	34	5	4
1	0,6	0,1	0,1
2	1,1	0,2	0,1
3	1,7	0,2	0,2
4	2,3	0,3	0,3
5	2,8	0,4	0,3
6	3,4	0,5	0,4
7	4,0	0,6	0,5
8	4,5	0,7	0,5
9	5,1	0,8	0,6
10	5,7	0,8	0,7
20	11,3	1,7	1,3
30	17,0	2,5	2,0
40	22,7	3,3	2,7
50	28,3	4,2	3,3

P. P.

70°

20°

'	L. Sin.	d.	L. Tang.	c.d.	L. Cotg.	L. Cos.	d.	'
0	9.53 405	35	9.56 107	39	0.43 893	9.97 299	5	60
1	9.53 440	35	9.56 146	39	0.43 854	9.97 294	5	59
2	9.53 475	34	9.56 185	39	0.43 815	9.97 289	4	58
3	9.53 509	35	9.56 224	40	0.43 776	9.97 285	5	57
4	9.53 544	34	9.56 264	39	0.43 736	9.97 280	4	56
5	9.53 578	35	9.56 303	39	0.43 697	9.97 276	5	55
6	9.53 613	34	9.56 342	39	0.43 658	9.97 271	5	54
7	9.53 647	35	9.56 381	39	0.43 619	9.97 266	4	53
8	9.53 682	34	9.56 420	39	0.43 580	9.97 262	5	52
9	9.53 716	35	9.56 459	39	0.43 541	9.97 257	5	51
10	9.53 751	34	9.56 498	39	0.43 502	9.97 252	5	50
11	9.53 785	34	9.56 537	39	0.43 463	9.97 248	5	49
12	9.53 819	35	9.56 576	39	0.43 424	9.97 243	5	48
13	9.53 854	34	9.56 615	39	0.43 385	9.97 238	4	47
14	9.53 888	34	9.56 654	39	0.43 346	9.97 234	5	46
15	9.53 922	35	9.56 693	39	0.43 307	9.97 229	5	45
16	9.53 957	34	9.56 732	39	0.43 268	9.97 224	4	44
17	9.53 991	34	9.56 771	39	0.43 229	9.97 220	5	43
18	9.54 025	34	9.56 810	39	0.43 190	9.97 215	5	42
19	9.54 059	34	9.56 849	38	0.43 151	9.97 210	4	41
20	9.54 093	34	9.56 887	39	0.43 113	9.97 206	5	40
21	9.54 127	34	9.56 926	39	0.43 074	9.97 201	5	39
22	9.54 161	34	9.56 965	39	0.43 035	9.97 196	4	38
23	9.54 195	34	9.57 004	38	0.42 996	9.97 192	5	37
24	9.54 229	34	9.57 042	39	0.42 958	9.97 187	5	36
25	9.54 263	34	9.57 081	39	0.42 919	9.97 182	4	35
26	9.54 297	34	9.57 120	38	0.42 880	9.97 178	5	34
27	9.54 331	34	9.57 158	39	0.42 842	9.97 173	5	33
28	9.54 365	34	9.57 197	38	0.42 803	9.97 168	5	32
29	9.54 399	34	9.57 235	39	0.42 765	9.97 163	4	31
30	9.54 433	33	9.57 274	38	0.42 726	9.97 159	5	30
31	9.54 466	34	9.57 312	39	0.42 688	9.97 154	5	29
32	9.54 500	34	9.57 351	38	0.42 649	9.97 149	4	28
33	9.54 534	33	9.57 389	39	0.42 611	9.97 145	5	27
34	9.54 567	34	9.57 428	38	0.42 572	9.97 140	5	26
35	9.54 601	34	9.57 466	38	0.42 534	9.97 135	5	25
36	9.54 635	33	9.57 504	39	0.42 496	9.97 130	4	24
37	9.54 668	34	9.57 543	38	0.42 457	9.97 126	5	23
38	9.54 702	33	9.57 581	38	0.42 419	9.97 121	5	22
39	9.54 735	34	9.57 619	39	0.42 381	9.97 116	5	21
40	9.54 769	33	9.57 658	38	0.42 342	9.97 111	4	20
41	9.54 802	34	9.57 696	38	0.42 304	9.97 107	5	19
42	9.54 836	33	9.57 734	38	0.42 266	9.97 102	5	18
43	9.54 869	34	9.57 772	38	0.42 228	9.97 097	5	17
44	9.54 903	33	9.57 810	38	0.42 190	9.97 092	5	16
45	9.54 936	33	9.57 849	38	0.42 151	9.97 087	4	15
46	9.54 969	34	9.57 887	38	0.42 113	9.97 083	5	14
47	9.55 003	33	9.57 925	38	0.42 075	9.97 078	5	13
48	9.55 036	33	9.57 963	38	0.42 037	9.97 073	5	12
49	9.55 069	33	9.58 001	38	0.41 999	9.97 068	5	11
50	9.55 102	34	9.58 039	38	0.41 961	9.97 063	4	10
51	9.55 136	33	9.58 077	38	0.41 923	9.97 059	5	9
52	9.55 169	33	9.58 115	38	0.41 885	9.97 054	5	8
53	9.55 202	33	9.58 153	38	0.41 847	9.97 049	5	7
54	9.55 235	33	9.58 191	38	0.41 809	9.97 044	5	6
55	9.55 268	33	9.58 229	38	0.41 771	9.97 039	4	5
56	9.55 301	33	9.58 267	37	0.41 733	9.97 035	5	4
57	9.55 334	33	9.58 304	38	0.41 696	9.97 030	5	3
58	9.55 367	33	9.58 342	38	0.41 658	9.97 025	5	2
59	9.55 400	33	9.58 380	38	0.41 620	9.97 020	5	1
60	9.55 433		9.58 418		0.41 582	9.97 015		0
	L. Cos.	d.	L. Cotg.	c.d.	L. Tang.	L. Sin.	d.	'

P. P.

	40	39	38
1	0,7	0,6	0,6
2	1,3	1,3	1,3
3	2,0	2,0	1,9
4	2,7	2,6	2,5
5	3,3	3,2	3,2
6	4,0	3,9	3,8
7	4,7	4,6	4,4
8	5,3	5,2	5,1
9	6,0	5,8	5,7
10	6,7	6,5	6,3
20	13,3	13,0	12,7
30	20,0	19,5	19,0
40	26,7	26,0	25,3
50	33,3	32,5	31,7

	37	35	34
1	0,6	0,6	0,6
2	1,2	1,2	1,1
3	1,8	1,8	1,7
4	2,5	2,3	2,3
5	3,1	2,9	2,8
6	3,7	3,5	3,4
7	4,3	4,1	4,0
8	4,9	4,7	4,5
9	5,6	5,2	5,1
10	6,2	5,8	5,7
20	12,3	11,7	11,3
30	18,5	17,5	17,0
40	24,7	23,3	22,7
50	30,8	29,2	28,3

	33	5	4
1	0,6	0,1	0,1
2	1,1	0,2	0,1
3	1,6	0,2	0,2
4	2,2	0,3	0,3
5	2,8	0,4	0,3
6	3,3	0,5	0,4
7	3,8	0,6	0,5
8	4,4	0,7	0,5
9	5,0	0,8	0,6
10	5,5	0,8	0,7
20	11,0	1,7	1,3
30	16,5	2,5	2,0
40	22,0	3,3	2,7
50	27,5	4,2	3,3

21°

'	L. Sin.	d.	L. Tang.	c.d.	L. Cotg.	L. Cos.	d.		P. P.		
0	9.55 433	88	9.58 418	87	0.41 582	9.97 015	5	60			
1	9.55 466	88	9.58 455	88	0.41 545	9.97 010	5	59			
2	9.55 499	88	9.58 498	88	0.41 507	9.97 005	4	58			
3	9.55 532	32	9.58 531	88	0.41 469	9.97 001	5	57		88	87	36
4	9.55 564	88	9.58 569	87	0.41 431	9.96 996	5	56			
5	9.55 597	88	9.58 606	88	0.41 394	9.96 991	5	55	1	0,6	0,6	0,6
6	9.55 630	88	9.58 644	87	0.41 356	9.96 986	5	54	2	1,8	1,2	1,2
7	9.55 663	32	9.58 681	88	0.41 319	9.96 981	5	58	8	1,9	1,8	1,8
8	9.55 695	88	9.58 719	88	0.41 281	9.96 976	5	52	4	2,5	2,5	2,4
9	9.55 728	88	9.58 757	87	0.41 243	9.96 971	5	51			
10	9.55 761	32	9.58 794	88	0.41 206	9.96 966	4	50	5	8,2	8,1	8,0
11	9.55 793	38	9.58 832	87	0.41 168	9.96 962	5	49	6	8,8	8,7	8,6
12	9.55 826	32	9.58 869	88	0.41 131	9.96 957	5	48	7	4,4	4,8	4,2
13	9.55 858	88	9.58 907	87	0.41 098	9.96 952	5	47	8	5,1	4,9	4,8
14	9.55 891	82	9.58 944	87	0.41 056	9.96 947	5	46	9	5,7	5,6	5,4
15	9.55 923	88	9.58 981	88	0.41 019	9.96 942	5	45	10	6,8	6,2	6,0
16	9.55 956	82	9.59 019	87	0.40 981	9.96 987	5	44	20	12,7	12,8	12,0
17	9.55 988	88	9.59 056	88	0.40 944	9.96 982	5	48	80	19,0	18,5	18,0
18	9.56 021	32	9.59 094	87	0.40 906	9.96 927	5	42	40	25,8	24,7	24,0
19	9.56 053	32	9.59 181	87	0.40 869	9.96 922	5	41	50	81,7	80,8	80,0
20	9.56 085	88	9.59 168	87	0.40 882	9.96 917	5	40			
21	9.56 118	32	9.59 205	88	0.40 795	9.96 912	5	39			
22	9.56 150	32	9.59 243	87	0.40 757	9.96 907	4	88			
28	9.56 182	88	9.59 280	87	0.40 720	9.96 908	5	87		88	32	81
24	9.56 215	32	9.59 317	87	0.40 688	9.96 898	5	86			
25	9.56 247	82	9.59 354	87	0.40 646	9.96 898	5	85	1	0,6	0,5	0,5
26	9.56 279	82	9.59 391	88	0.40 609	9.96 888	5	84	2	1,1	1,1	1,0
27	9.56 811	82	9.59 429	87	0.40 571	9.96 888	5	88	8	1,6	1,6	1,6
28	9.56 843	82	9.59 466	87	0.40 584	9.96 878	5	82	4	2,2	2,1	2,1
29	9.56 375	88	9.59 503	87	0.40 497	9.96 878	5	81	5	2,8	2,7	2,6
30	9.56 408	82	9.59 540	87	0.40 460	9.96 868	5	80	6	8,8	8,2	8,1
81	9.56 440	32	9.59 577	87	0.40 428	9.96 863	5	29	7	8,8	8,7	8,6
82	9.56 472	82	9.59 614	87	0.40 886	9.96 858	5	28	8	4,4	4,8	4,1
88	9.56 504	82	9.59 651	87	0.40 349	9.96 858	5	27	9	5,0	4,8	4,6
84	9.56 536	32	9.59 688	87	0.40 312	9.96 848	5	26			
85	9.56 568	81	9.59 725	87	0.40 275	9.96 848	5	25	10	5,5	5,8	5,2
86	9.56 599	32	9.59 762	87	0.40 288	9.96 888	5	24	20	11,0	10,7	10,8
87	9.56 631	32	9.59 799	86	0.40 201	9.96 888	5	28	30	16,5	16,0	15,5
88	9.56 663	32	9.59 835	87	0.40 165	9.96 828	5	22	40	22,0	21,8	20,7
89	9.56 695	32	9.59 872	87	0.40 128	9.96 828	5	21	50	27,5	26,7	25,8
40	9.56 727	32	9.59 909	87	0.40 091	9.96 818	5	20			
41	9.56 759	81	9.59 946	87	0.40 054	9.96 818	5	19			
42	9.56 790	32	9.59 988	86	0.40 017	9.96 808	5	18			
48	9.56 822	32	9.60 019	87	0.39 981	9.96 808	5	17		6	5	4
44	9.56 854	32	9.60 056	87	0.39 944	9.96 798	5	16			
45	9.56 886	81	9.60 093	87	0.39 907	9.96 798	5	15	1	0,1	0,1	0,1
46	9.56 917	32	9.60 130	86	0.39 870	9.96 788	5	14	2	0,2	0,2	0,1
47	9.56 949	81	9.60 166	87	0.39 884	9.96 788	5	18	8	0,8	0,2	0,2
48	9.56 980	32	9.60 208	87	0.39 797	9.96 778	6	12	4	0,4	0,8	0,3
49	9.57 012	32	9.60 240	86	0.39 760	9.96 772	5	11	5	0,5	0,4	0,8
50	9.57 044	81	9.60 276	87	0.39 724	9.96 767	5	10	6	0,6	0,5	0,4
51	9.57 075	82	9.60 818	86	0.39 687	9.96 762	5	9	7	0,7	0,6	0,5
52	9.57 107	81	9.60 349	87	0.39 651	9.96 757	5	8	8	0,8	0,7	0,5
58	9.57 188	81	9.60 886	86	0.39 614	9.96 752	5	7	9	0,9	0,8	0,6
54	9.57 169	82	9.60 422	87	0.39 578	9.96 747	5	6			
55	9.57 201	81	9.60 459	86	0.39 541	9.96 742	5	5	10	1,0	0,8	0,7
56	9.57 232	82	9.60 495	87	0.39 505	9.96 787	5	4	80	2,0	1,7	1,8
57	9.57 264	81	9.60 568	86	0.39 468	9.96 782	5	8	80	8,0	2,5	2,0
58	9.57 295	81	9.60 568	87	0.39 432	9.96 727	5	2	40	4,0	8,8	2,7
59	9.57 826	82	9.60 605	86	0.39 895	9.96 722	5	1	50	5,0	4,2	8,8
60	9.57 858		9.60 641		0.39 359	9.96 717		0			
	L. Cos.	d.	L.Cotg.	c.d.	L. Tang.	L. Sin.	d.	'		P. P.	

68°

′	L. Sin.	d.	L. Tang.	c.d.	L. Cotg.	L. Cos.	d.	′
0	9.57 358	31	9.60 641	36	0.39 359	9.96 717	6	60
1	9.57 389	31	9.60 677	37	0.39 323	9.96 711	5	59
2	9.57 420	31	9.60 714	36	0.39 286	9.96 706	5	58
3	9.57 451	31	9.60 750	36	0.39 250	9.96 701	5	57
4	9.57 482	32	9.60 786	37	0.39 214	9.96 696	5	56
5	9.57 514	31	9.60 823	36	0.39 177	9.96 691	5	55
6	9.57 545	31	9.60 859	36	0.39 141	9.96 686	5	54
7	9.57 576	31	9.60 895	36	0.39 105	9.96 631	5	53
8	9.57 607	31	9.60 931	36	0.39 069	9.96 676	6	52
9	9.57 638	31	9.60 967	37	0.39 033	9.96 670	5	51
10	9.57 669	31	9.61 004	36	0.38 996	9.96 665	5	50
11	9.57 700	31	9.61 040	36	0.38 960	9.96 660	5	49
12	9.57 731	31	9.61 076	36	0.38 924	9.96 655	5	48
13	9.57 762	31	9.61 112	36	0.38 888	9.96 650	5	47
14	9.57 793	31	9.61 148	36	0.38 852	9.96 645	5	46
15	9.57 824	31	9.61 184	36	0.38 816	9.96 640	6	45
16	9.57 855	30	9.61 220	36	0.38 780	9.96 634	5	44
17	9.57 885	31	9.61 256	36	0.38 744	9.96 629	5	43
18	9.57 916	31	9.61 292	36	0.38 708	9.96 624	5	42
19	9.57 947	31	9.61 328	36	0.38 672	9.96 619	5	41
20	9.57 978	30	9.61 364	36	0.38 636	9.96 614	6	40
21	9.58 008	31	9.61 400	36	0.38 600	9.96 608	5	39
22	9.58 039	31	9.61 436	36	0.38 564	9.96 603	5	38
23	9.58 070	31	9.61 472	36	0.38 528	9.96 598	5	37
24	9.58 101	30	9.61 508	36	0.38 492	9.96 593	5	36
25	9.58 131	31	9.61 544	35	0.38 456	9.96 588	6	35
26	9.58 162	30	9.61 579	36	0.38 421	9.96 582	5	34
27	9.58 192	31	9.61 615	36	0.38 385	9.96 577	5	33
28	9.58 223	30	9.61 651	36	0.38 349	9.96 572	5	32
29	9.58 253	31	9.61 687	35	0.38 313	9.96 567	5	31
30	9.58 284	30	9.61 722	36	0.38 278	9.96 562	6	30
31	9.58 314	31	9.61 758	36	0.38 242	9.96 556	5	29
32	9.58 345	30	9.61 794	36	0.38 206	9.96 551	5	28
33	9.58 375	31	9.61 830	35	0.38 170	9.96 546	5	27
34	9.58 406	30	9.61 865	36	0.38 135	9.96 541	6	26
35	9.58 436	31	9.61 901	35	0.38 099	9.96 535	5	25
36	9.58 467	30	9.61 936	36	0.38 064	9.96 530	5	24
37	9.58 497	30	9.61 972	36	0.38 028	9.96 525	5	23
38	9.58 527	30	9.62 008	35	0.37 992	9.96 520	6	22
39	9.58 557	31	9.62 043	36	0.37 957	9.96 514	5	21
40	9.58 588	30	9.62 079	35	0.37 921	9.96 509	5	20
41	9.58 618	30	9.62 114	36	0.37 886	9.96 504	6	19
42	9.58 648	30	9.62 150	35	0.37 850	9.96 498	5	18
43	9.58 678	31	9.62 185	36	0.37 815	9.96 493	5	17
44	9.58 709	30	9.62 221	35	0.37 779	9.96 488	5	16
45	9.58 739	30	9.62 256	36	0.37 744	9.96 483	6	15
46	9.58 769	30	9.62 292	35	0.37 708	9.96 477	5	14
47	9.58 799	30	9.62 327	35	0.37 673	9.96 472	5	13
48	9.58 829	30	9.62 362	36	0.37 638	9.96 467	6	12
49	9.58 859	30	9.62 398	35	0.37 602	9.96 461	5	11
50	9.58 889	30	9.62 433	35	0.37 567	9.96 456	5	10
51	9.58 919	30	9.62 468	36	0.37 532	9.96 451	5	9
52	9.58 949	30	9.62 504	35	0.37 496	9.96 445	5	8
53	9.58 979	30	9.62 539	35	0.37 461	9.96 440	5	7
54	9.59 009	30	9.62 574	35	0.37 426	9.96 435	6	6
55	9.59 039	30	9.62 609	36	0.37 391	9.96 429	5	5
56	9.59 069	29	9.62 645	35	0.37 355	9.96 424	5	4
57	9.59 098	30	9.62 680	35	0.37 320	9.96 419	6	3
58	9.59 128	30	9.62 715	35	0.37 285	9.96 413	5	2
59	9.59 158	30	9.62 750	35	0.37 250	9.96 408	5	1
60	9.59 188		9.62 785		0.37 215	9.96 403		0

| | L. Cos. | d. | L.Cotg. | c.d. | L. Tang. | L. Sin. | d. | ′ |

P. P.

	37	36	35
1	0,6	0,6	0,6
2	1,2	1,2	1,2
3	1,8	1,8	1,8
4	2,5	2,4	2,3
5	3,1	3,0	2,9
6	3,7	3,6	3,5
7	4,3	4,2	4,1
8	4,9	4,8	4,7
9	5,6	5,4	5,2
10	6,2	6,0	5,8
20	12,3	12,0	11,7
30	18,5	18,0	17,5
40	24,7	24,0	23,3
50	30,8	30,0	29,2

	32	31	30
1	0,5	0,5	0,5
2	1,1	1,0	1,0
3	1,6	1,6	1,5
4	2,1	2,1	2,0
5	2,7	2,6	2,5
6	3,2	3,1	3,0
7	3,7	3,6	3,5
8	4,3	4,1	4,0
9	4,8	4,6	4,5
10	5,3	5,2	5,0
20	10,7	10,3	10,0
30	16,0	15,5	15,0
40	21,3	20,7	20,0
50	26,7	25,8	25,0

	29	6	5
1	0,5	0,1	0,1
2	1,0	0,2	0,2
3	1,4	0,3	0,2
4	1,9	0,4	0,3
5	2,4	0,5	0,4
6	2,9	0,6	0,5
7	3,4	0,7	0,6
8	3,9	0,8	0,7
9	4,4	0,9	0,8
10	4,8	1,0	0,8
20	9,7	2,0	1,7
30	14,5	3,0	2,5
40	19,3	4,0	3,3
50	24,2	5,0	4,2

23°

'	L. Sin.	d.	L. Tang.	c.d.	L. Cotg.	L. Cos.	d.		P. P.		
0	9.59 188	30	9.62 785	35	0.37 215	9.96 408	6	60			
1	9.59 218	29	9.62 820	35	0.37 180	9.96 397	5	59			
2	9.59 247	30	.62 855	35	0.37 145	9.96 392	5	58			
3	9.59 277	30	.62 890	36	0.37 110	9.96 387	6	57		**36**	**35**	**34**
4	9.59 307	29	.62 926	35	0.37 074	9.96 381	5	56			
5	9.59 336	30	9.62 961	35	0.37 039	9.96 376	6	55	1	0,6	0,6	0,6
6	9.59 366	30	9.62 996	35	0.37 004	9.96 370	5	54	2	1,2	1,2	1,1
7	59 396	29	.63 031	35	0.36 969	9.96 365	5	53	3	1,8	1,8	1,7
8	9.59 425	30	.63 066	35	0.36 934	9.96 360	6	52	4	2,4	2,3	2,3
9	9.59 455	29	.63 101	34	0.36 899	9.96 354	5	51			
10	9.59 484	30	9.63 135	35	0.36 865	9.96 349	6	50	5	3,0	2,9	2,8
11	9.59 514	30	9.63 170	35	0.36 830	9.96 343	6	49	6	3,6	3,5	3,4
12	9.59 548	30	9.63 205	35	0.36 795	9.96 338	5	48	7	4,2	4,1	4,0
13	.59 573	29	.63 240	35	0.36 760	9.96 333	5	47	8	4,8	4,7	4,5
14	.59 602	30	.63 275	35	0.36 725	9.96 327	6	46	9	5,4	5,2	5,1
15	9.59 632	29	9.63 310	35	0.36 690	9.96 322	6	45			
16	9.59 661	29	9.63 345	34	0.36 655	9.96 316	5	44	10	6,0	5,8	5,7
17	.59 690	30	.63 379	35	0.36 621	9.96 311	6	43	20	12,0	11,7	11,3
18	.59 720	29	.63 414	35	0.36 586	9.96 305	5	42	30	18,0	17,5	17,0
19	.59 749	29	.63 449	35	0.36 551	9.96 300	6	41	40	24,0	23,3	22,7
20	9.59 778	30	9.63 484	35	0.36 516	9.96 294	5	40	50	30,0	29,2	28,3
21	9.59 806	29	9.63 519	34	0.36 481	9.96 289	5	39			
22	59 837	29	.63 553	35	0.36 447	9.96 284	5	38			
23	59 866	29	.63 588	35	0.36 412	9.96 278	5	37		**30**	**29**	**28**
24	59 895	29	.63 623	34	0.36 377	9.96 273	6	36			
25	9 59 924	30	9.63 657	35	0.36 343	9.96 267	5	35	1	0,5	0,5	0,5
26	9 59 954	29	9.63 692	34	0.36 308	9.96 262	6	34	2	1,0	1,0	0,9
27	59 983	29	.63 726	35	0.36 274	9.96 256	5	33	3	1,5	1,4	1,4
28	60 012	29	.63 761	35	0.36 239	9.96 251	6	32	4	2,0	1,9	1,9
29	60 041	29	.63 796	34	0.36 204	9.96 245	5	31			
30	9.60 070	29	9.63 830	35	0.36 170	9.96 240	6	30	5	2,5	2,4	2,3
31	9.60 099	29	9.63 865	34	0.36 135	9.96 234	5	29	6	3,0	2,9	2,8
32	60 128	29	.63 899	35	0.36 101	9.96 229	5	28	7	3,5	3,4	3,3
33	60 157	29	.63 934	34	0.36 066	9.96 223	5	27	8	4,0	3,9	3,7
34	60 186	29	.63 968	35	0.36 032	9.96 218	6	26	9	4,5	4,4	4,2
35	9 60 215	29	9.64 003	34	0.35 997	9.96 212	5	25			
36	9 60 244	29	9.64 037	35	0.35 963	9.96 207	5	24	10	5,0	4,8	4,7
37	60 273	29	.64 072	34	0.35 928	9.96 201	5	23	20	10,0	9,7	9,3
38	60 302	29	.64 106	34	0.35 894	9.96 196	6	22	30	15,0	14,5	14,0
39	60 331	28	.64 140	35	0.35 860	9.96 190	5	21	40	20,0	19,3	18,7
40	9 60 359	29	9.64 175	34	0.35 825	9.96 185	6	20	50	25,0	24,2	23,3
41	9 60 388	29	9.64 209	34	0.35 791	9.96 179	5	19			
42	60 417	29	.64 243	35	0.35 757	9.96 174	6	18			
43	60 446	28	.64 278	34	0.35 722	9.96 168	6	17		**6**	**5**	
44	60 474	29	.64 312	34	0.35 688	9.96 162	5	16			
45	9 60 503	29	9.64 346	35	0.35 654	9.96 157	6	15	1	0,1	0,1	
46	9 60 532	29	9.64 381	34	0.35 619	9.96 151	5	14	2	0,2	0,2	
47	60 561	28	.64 415	34	.35 585	9.96 146	6	13	3	0,3	0,2	
48	60 589	29	.64 449	34	.35 551	9.96 140	5	12	4	0,4	0,3	
49	60 618	28	.64 483	34	.35 517	9.96 135	6	11			
50	9.60 646	29	9.64 517	35	0.35 483	9.96 129	6	10	5	0,5	0,4	
51	9.60 675	29	9.64 552	34	0.35 448	9.96 123	5	9	6	0,6	0,5	
52	60 704	28	64 586	34	.35 414	9.96 118	6	8	7	0,7	0,6	
53	9.60 732	29	64 620	34	.35 380	9.96 112	5	7	8	0,8	0,7	
54	.60 761	28	64 654	34	.35 346	9.96 107	6	6	9	0,9	0,8	
55	9.60 789	29	9.64 688	34	0.35 312	9.96 101	6	5			
56	9.60 818	28	9.64 722	34	0.35 278	9.96 095	5	4	10	1,0	0,8	
57	.60 846	29	.64 756	34	.35 244	9.96 090	6	3	20	2,0	1,7	
58	.60 875	28	.64 790	34	.35 210	9.96 084	5	2	30	3,0	2,5	
59	9.60 903	28	64 824	34	.35 176	9.96 079	6	1	40	4,0	3,3	
60	9.60 931		9.64 858		0.35 142	9.96 073		0	50	5,0	4,2	

| | L. Cos. | d. | L. Cotg. | c.d. | L. Tang. | L. Sin. | d. | ' | | P. P. | |

66°

24°

'	L. Sin.	d.	L. Tang.	c.d.	L. Cotg.	L. Cos.	d.		P. P.
0	9.60 981		9.64 858		0.35 142	9.96 078		60	
1	9.60 960	29	9.64 892	34	0.35 108	9.96 067	6	59	
2	9.60 988	28	9.64 926	34	0.35 074	9.96 062	5	58	
3	9.61 016	28	9.64 960	34	0.35 040	9.96 056	6	57	
4	9.61 045	29	9.64 994	34	0.35 006	9.96 050	5	56	**34** **33**
		28		34			6		
5	9.61 073	28	9.65 028	34	0.34 972	9.96 045		55	1 0,6 0,6
6	9.61 101	28	9.65 062	34	0.34 938	9.96 039	6	54	2 1,1 1,1
7	9.61 129	29	9.65 096	34	0.34 904	9.96 034	5	53	3 1,7 1,6
8	9.61 158	28	9.65 130	34	0.34 870	9.96 028	6	52	4 2,3 2,2
9	9.61 186	28	9.65 164	33	0.34 836	9.96 022	5	51	
10	9.61 214	28	9.65 197	34	0.34 803	9.96 017	6	50	5 2,8 2,8
11	9.61 242	28	9.65 231	34	0.34 769	9.96 011	6	49	6 3,4 3,3
12	9.61 270	28	9.65 265	34	0.34 735	9.96 005	5	48	7 4,0 3,8
13	9.61 298	28	9.65 299	34	0.34 701	9.96 000	6	47	8 4,5 4,4
14	9.61 326	28	9.65 333	33	0.34 667	9.95 994	6	46	9 5,1 5,0
15	9.61 354	28	9.65 366	34	0.34 634	9.95 988	6	45	10 5,7 5,5
16	9.61 382	29	9.65 400	34	0.34 600	9.95 982	5	44	20 11,3 11,0
17	9.61 411	27	9.65 434	33	0.34 566	9.95 977	6	43	30 17,0 16,5
18	9.61 438	28	9.65 467	34	0.34 533	9.95 971	6	42	40 22,7 22,0
19	9.61 466	28	9.65 501	34	0.34 499	9.95 965	5	41	50 28,3 27,5
20	9.61 494	28	9.65 535	33	0.34 465	9.95 960	6	40	
21	9.61 522	28	9.65 568	34	0.34 432	9.95 954	6	39	
22	9.61 550	28	9.65 602	34	0.34 398	9.95 948	6	38	
23	9.61 578	28	9.65 636	33	0.34 364	9.95 942	6	37	
24	9.61 606	28	9.65 669	34	0.34 331	9.95 937	5	36	**29** **28** **27**
25	9.61 634	28	9.65 703	33	0.34 297	9.95 931	6	35	1 0,5 0,5 0,4
26	9.61 662	27	9.65 736	34	0.34 264	9.95 925	5	34	2 1,0 0,9 0,9
27	9.61 689	28	9.65 770	33	0.34 230	9.95 920	6	33	3 1,4 1,4 1,4
28	9.61 717	28	9.65 803	34	0.34 197	9.95 914	6	32	4 1,9 1,9 1,8
29	9.61 745	28	9.65 837	33	0.34 163	9.95 908	6	31	
30	9.61 773	27	9.65 870	34	0.34 130	9.95 902	5	30	5 2,4 2,3 2,2
31	9.61 800	28	9.65 904	33	0.34 096	9.95 897	6	29	6 2,9 2,8 2,7
32	9.61 828	28	9.65 937	34	0.34 063	9.95 891	6	28	7 3,4 3,3 3,2
33	9.61 856	27	9.65 971	33	0.34 029	9.95 885	6	27	8 3,9 3,7 3,6
34	9.61 883	28	9.66 004	34	0.33 996	9.95 879	6	26	9 4,4 4,2 4,0
35	9.61 911	28	9.66 038	33	0.33 962	9.95 873	5	25	10 4,8 4,7 4,5
36	9.61 939	27	9.66 071	33	0.33 929	9.95 868	6	24	20 9,7 9,3 9,0
37	9.61 966	28	9.66 104	34	0.33 896	9.95 862	6	23	30 14,5 14,0 13,5
38	9.61 994	27	9.66 138	33	0.33 862	9.95 856	6	22	40 19,3 18,7 18,0
39	9.62 021	28	9.66 171	33	0.33 829	9.95 850	6	21	50 24,2 23,3 22,5
40	9.62 049	27	9.66 204	34	0.33 796	9.95 844	5	20	
41	9.62 076	28	9.66 238	33	0.33 762	9.95 839	6	19	
42	9.62 104	27	9.66 271	33	0.33 729	9.95 833	6	18	
43	9.62 131	28	9.66 304	33	0.33 696	9.95 827	6	17	**6** **5**
44	9.62 159	27	9.66 337	34	0.33 663	9.95 821	6	16	1 0,1 0,1
45	9.62 186	28	9.66 371	33	0.33 629	9.95 815	5	15	2 0,2 0,2
46	9.62 214	27	9.66 404	33	0.33 596	9.95 810	6	14	3 0,3 0,2
47	9.62 241	27	9.66 437	33	0.33 563	9.95 804	6	13	4 0,4 0,3
48	9.62 268	28	9.66 470	33	0.33 530	9.95 798	6	12	
49	9.62 296	27	9.66 503	34	0.33 497	9.95 792	6	11	5 0,5 0,4
50	9.62 323	27	9.66 537	33	0.33 463	9.95 786	6	10	6 0,6 0,5
51	9.62 350	27	9.66 570	33	0.33 430	9.95 780	5	9	7 0,7 0,6
52	9.62 377	28	9.66 603	33	0.33 397	9.95 775	6	8	8 0,8 0,7
53	9.62 405	27	9.66 636	33	0.33 364	9.95 769	6	7	9 0,9 0,8
54	9.62 432	27	9.66 669	33	0.33 331	9.95 763	6	6	
55	9.62 459	27	9.66 702	33	0.33 298	9.95 757	6	5	10 1,0 0,8
56	9.62 486	27	9.66 735	33	0.33 265	9.95 751	6	4	20 2,0 1,7
57	9.62 513	28	9.66 768	33	0.33 232	9.95 745	6	3	30 3,0 2,5
58	9.62 541	27	9.66 801	33	0.33 199	9.95 739	6	2	40 4,0 3,3
59	9.62 568	27	9.66 834	33	0.33 166	9.95 733	5	1	50 5,0 4,2
60	9.62 595		9.66 867		0.33 133	9.95 728		0	
	L. Cos.	d.	L. Cotg.	c.d.	L. Tang.	L. Sin.	d.	'	P. P.

65°

'	L. Sin.	d.	L. Tang.	c.d.	L. Cotg.	L. Cos.	d.		P. P.		
0	9.62 595	27	9.66 867	33	0.33 133	9.95 728	6	60			
1	9.62 622	27	9.66 900	33	0.33 100	9.95 722	6	59			
2	9.62 649	27	9.66 933	33	0.33 067	9.95 716	6	58			
3	9.62 676	27	9.66 966	33	0.33 034	9.95 710	6	57			
4	9.62 703	27	9.66 999	33	0.33 001	9.95 704	6	56		**33**	**32**
5	9.62 730	27	9.67 032	33	0.32 968	9.95 698	6	55	1	0,6	0,5
6	9.62 757	27	9.67 065	33	0.32 935	9.95 692	6	54	2	1,1	1,1
7	9.62 784	27	9.67 098	33	0.32 902	9.95 686	6	53	3	1,6	1,6
8	9.62 811	27	9.67 131	32	0.32 869	9.95 680	6	52	4	2,2	2,1
9	9.62 838	27	9.67 163	33	0.32 837	9.95 674	6	51			
10	9.62 865	27	9.67 196	33	0.32 804	9.95 668	5	50	5	2,8	2,7
11	9.62 892	26	9.67 229	33	0.32 771	9.95 663	6	49	6	3,3	3,2
12	9.62 918	27	9.67 262	33	0.32 738	9.95 657	6	48	7	3,8	3,7
13	9.62 945	27	9.67 295	32	0.32 705	9.95 651	6	47	8	4,4	4,3
14	9.62 972	27	9.67 327	33	0.32 673	9.95 645	6	46	9	5,0	4,8
15	9.62 999	27	9.67 360	33	0.32 640	9.95 639	6	45			
16	9.63 026	26	9.67 393	33	0.32 607	9.95 633	6	44	10	5,5	5,3
17	9.63 052	27	9.67 426	32	0.32 574	9.95 627	6	43	20	11,0	10,7
18	9.63 079	27	9.67 458	33	0.32 542	9.95 621	6	42	30	16,5	16,0
19	9.63 106	27	9.67 491	33	0.32 509	9.95 615	6	41	40	22,0	21,3
20	9.63 133	26	9.67 524	32	0.32 476	9.95 609	6	40	50	27,5	26,7
21	9.63 159	27	9.67 556	33	0.32 444	9.95 603	6	39			
22	9.63 186	27	9.67 589	33	0.32 411	9.95 597	6	38			
23	9.63 213	26	9.67 622	32	0.32 378	9.95 591	6	37		**27**	**26**
24	9.63 239	27	9.67 654	33	0.32 346	9.95 585	6	36			
25	9.63 266	26	9.67 687	32	0.32 313	9.95 579	6	35	1	0,4	0,4
26	9.63 292	27	9.67 719	33	0.32 281	9.95 573	6	34	2	0,9	0,9
27	9.63 319	26	9.67 752	33	0.32 248	9.95 567	6	33	3	1,4	1,3
28	9.63 345	27	9.67 785	32	0.32 215	9.95 561	6	32	4	1,8	1,7
29	9.63 372	26	9.67 817	33	0.32 183	9.95 555	6	31			
30	9.63 398	27	9.67 850	32	0.32 150	9.95 549	6	30	5	2,2	2,2
31	9.63 425	26	9.67 882	33	0.32 118	9.95 543	6	29	6	2,7	2,6
32	9.63 451	27	9.67 915	32	0.32 085	9.95 537	6	28	7	3,2	3,0
33	9.63 478	26	9.67 947	33	0.32 053	9.95 531	6	27	8	3,6	3,5
34	9.63 504	27	9.67 980	32	0.32 020	9.95 525	6	26	9	4,0	3,9
35	9.63 531	26	9.68 012	32	0.31 988	9.95 519	6	25			
36	9.63 557	26	9.68 044	33	0.31 956	9.95 513	6	24	10	4,5	4,3
37	9.63 583	27	9.68 077	32	0.31 923	9.95 507	7	23	20	9,0	8,7
38	9.63 610	26	9.68 109	33	0.31 891	9.95 500	6	22	30	13,5	13,0
39	9.63 636	26	9.68 142	32	0.31 858	9.95 494	6	21	40	18,0	17,3
40	9.63 662	27	9.68 174	32	0.31 826	9.95 488	6	20	50	22,5	21,7
41	9.63 689	26	9.68 206	33	0.31 794	9.95 482	6	19			
42	9.63 715	26	9.68 239	32	0.31 761	9.95 476	6	18			
43	9.63 741	26	9.68 271	32	0.31 729	9.95 470	6	17		**7**	**6**	**5**
44	9.63 767	27	9.68 303	33	0.31 697	9.95 464	6	16				
45	9.63 794	26	9.68 336	32	0.31 664	9.95 458	6	15	1	0,1	0,1	0,1
46	9.63 820	26	9.68 368	32	0.31 632	9.95 452	6	14	2	0,2	0,2	0,2
47	9.63 846	26	9.68 400	32	0.31 600	9.95 446	6	13	3	0,4	0,3	0,2
48	9.63 872	26	9.68 432	33	0.31 568	9.95 440	6	12	4	0,5	0,4	0,3
49	9.63 898	26	9.68 465	32	0.31 535	9.95 434	7	11				
50	9.63 924	26	9.68 497	32	0.31 503	9.95 427	6	10	5	0,6	0,5	0,4
									6	0,7	0,6	0,5
51	9.63 950	26	9.68 529	32	0.31 471	9.95 421	6	9	7	0,8	0,7	0,6
52	9.63 976	26	9.68 561	32	0.31 439	9.95 415	6	8	8	0,9	0,8	0,7
53	9.64 002	26	9.68 593	33	0.31 407	9.95 409	6	7	9	1,0	0,9	0,8
54	9.64 028	26	9.68 626	32	0.31 374	9.95 403	6	6				
55	9.64 054	26	9.68 658	32	0.31 342	9.95 397	6	5	10	1,2	1,0	0,8
56	9.64 080	26	9.68 690	32	0.31 310	9.95 391	7	4	20	2,3	2,0	1,7
57	9.64 106	26	9.68 722	32	0.31 278	9.95 384	6	3	30	3,5	3,0	2,5
58	9.64 132	26	9.68 754	32	0.31 246	9.95 378	6	2	40	4,7	4,0	3,3
59	9.64 158	26	9.68 786	32	0.31 214	9.95 372	6	1	50	5,8	5,0	4,2
60	9.64 184		9.68 818		0.31 182	9.95 366		0				

| | L. Cos. | d. | L.Cotg. | c.d. | L. Tang. | L. Sin. | d. | ' | P. P. | | |

'	L. Sin.	d.	L. Tang.	c.d.	L. Cotg.	L. Cos.	d.	
0	9.64 184	26	9.68 818	32	0.81 182	9.95 366		60
1	9.64 210	26	9.68 850	32	0.81 150	9.95 360	6	59
2	9.64 236	26	9.68 882	32	0.81 118	9.95 354	6	58
3	9.64 262	26	9.68 914	32	0.81 086	9.95 348	6	57
4	9.64 288	25	9.68 946	32	0.81 054	9.95 341	7 6	56
5	9.64 313	26	9.68 978	32	0.81 022	9.95 335	6	55
6	9.64 339	26	9.69 010	32	0.80 990	9.95 329	6	54
7	9.64 365	26	9.69 042	32	0.80 958	9.95 323	6	53
8	9.64 391	26	9.69 074	32	0.80 926	9.95 317	7 6	52
9	9.64 417	25	9.69 106	32	0.80 894	9.95 310		51
10	9.64 442	26	9.69 138	32	0.80 862	9.95 304	6	50
11	9.64 468	26	9.69 170	32	0.80 830	9.95 298	6	49
12	9.64 494	25	9.69 202	32	0.80 798	9.95 292	6	48
13	9.64 519	26	9.69 234	32	0.80 766	9.95 286	7 6	47
14	9.64 545	26	9.69 266	32	0.80 734	9.95 279		46
15	9.64 571	25	9.69 298	31	0.80 702	9.95 273	6	45
16	9.64 596	26	9.69 329	32	0.80 671	9.95 267	6	44
17	9.64 622	25	9.69 361	32	0.80 639	9.95 261	7 6	43
18	9.64 647	26	9.69 393	32	0.80 607	9.95 254		42
19	9.64 673	25	9.69 425	32	0.80 575	9.95 248	6	41
20	9.64 698	26	9.69 457	31	0.80 543	9.95 242	6	40
21	9.64 724	25	9.69 488	32	0.80 512	9.95 236	7	39
22	9.64 749	26	9.69 520	32	0.80 480	9.95 229	6	38
23	9.64 775	25	9.69 552	32	0.80 448	9.95 223	6	37
24	9.64 800	26	9.69 584	31	0.80 416	9.95 217	6	36
25	9.64 826	25	9.69 615	32	0.80 385	9.95 211	7	35
26	9.64 851	26	9.69 647	32	0.80 353	9.95 204	6	34
27	9.64 877	25	9.69 679	31	0.80 321	9.95 198	6	33
28	9.64 902	25	9.69 710	32	0.80 290	9.95 192	7 6	32
29	9.64 927	26	9.69 742	32	0.80 258	9.95 185		31
30	9.64 953	25	9.69 774	31	0.80 226	9.95 179	6	30
31	9.64 978	25	9.69 805	32	0.80 195	9.95 173	6	29
32	9.65 003	26	9.69 837	31	0.80 163	9.95 167	7 6	28
33	9.65 029	25	9.69 868	32	0.80 132	9.95 160		27
34	9.65 054	25	9.69 900	32	0.80 100	9.95 154	6	26
35	9.65 079	25	9.69 932	31	0.80 068	9.95 148	7	25
36	9.65 104	26	9.69 963	32	0.80 037	9.95 141	6	24
37	9.65 130	25	9.69 995	31	0.80 005	9.95 135	6	23
38	9.65 155	25	9.70 026	32	0.29 974	9.95 129	7 6	22
39	9.65 180	25	9.70 058	31	0.29 942	9.95 122		21
40	9.65 205	25	9.70 089	32	0.29 911	9.95 116	6	20
41	9.65 230	25	9.70 121	31	0.29 879	9.95 110	7	19
42	9.65 255	26	9.70 152	32	0.29 848	9.95 103	6	18
43	9.65 281	25	9.70 184	31	0.29 816	9.95 097	7 6	17
44	9.65 306	25	9.70 215	32	0.29 785	9.95 090		16
45	9.65 331	25	9.70 247	31	0.29 753	9.95 084	6	15
46	9.65 356	25	9.70 278	31	0.29 722	9.95 078	7	14
47	9.65 381	25	9.70 309	32	0.29 691	9.95 071	6	13
48	9.65 406	25	9.70 341	31	0.29 659	9.95 065	6	12
49	9.65 431	25	9.70 372	32	0.29 628	9.95 059	7	11
50	9.65 456	25	9.70 404	31	0.29 596	9.95 052	6	10
51	9.65 481	25	9.70 435	31	0.29 565	9.95 046	7	9
52	9.65 506	25	9.70 466	32	0.29 534	9.95 039	6	8
53	9.65 531	25	9.70 498	31	0.29 502	9.95 033	6	7
54	9.65 556	24	9.70 529	31	0.29 471	9.95 027	7	6
55	9.65 580	25	9.70 560	32	0.29 440	9.95 020	6	5
56	9.65 605	25	9.70 592	31	0.29 408	9.95 014	7	4
57	9.65 630	25	9.70 623	31	0.29 377	9.95 007	6	3
58	9.65 655	25	9.70 654	31	0.29 346	9.95 001	6	2
59	9.65 680	25	9.70 685	32	0.29 315	9.94 995	7	1
60	9.65 705		9.70 717		0.29 283	9.94 988		0
	L. Cos.	d.	L. Cotg.	c.d.	L. Tang.	L. Sin.	d.	'

P. P.

	32	31
1	0,5	0,5
2	1,1	1,0
3	1,6	1,6
4	2,1	2,1
5	2,7	2,6
6	3,2	3,1
7	3,7	3,6
8	4,3	4,1
9	4,8	4,6
10	5,8	5,2
20	10,7	10,3
30	16,0	15,5
40	21,3	20,7
50	26,7	25,8

	26	25	24
1	0,4	0,4	0,4
2	0,9	0,8	0,8
3	1,3	1,2	1,2
4	1,7	1,7	1,6
5	2,2	2,1	2,0
6	2,6	2,5	2,4
7	3,0	2,9	2,8
8	3,5	3,3	3,2
9	3,9	3,8	3,6
10	4,3	4,2	4,0
20	8,7	8,3	8,0
30	13,0	12,5	12,0
40	17,3	16,7	16,0
50	21,7	20,8	20,0

	7	6
1	0,1	0,1
2	0,2	0,2
3	0,4	0,3
4	0,5	0,4
5	0,6	0,5
6	0,7	0,6
7	0,8	0,7
8	0,9	0,8
9	1,0	0,9
10	1,2	1,0
20	2,3	2,0
30	3,5	3,0
40	4,7	4,0
50	5,8	5,0

′	L. Sin.	d.	L. Tang.	c.d.	L. Cotg.	L. Cos.	d.	′	P. P.		
0	9.65 705		9.70 717		0.29 283	9.94 988		60			
1	9.65 729	24	9.70 748	31	0.29 252	9.94 982	6	59			
2	9.65 754	25	9.70 779	31	0.29 221	9.94 975	7	58			
3	9.65 779	25	9.70 810	31	0.29 190	9.94 969	6	57			
4	9.65 804	25	9.70 841	31	0.29 159	9.94 962	7	56		**32**	**31**	**30**
5	9.65 828	24	9.70 873	32	0.29 127	9.94 956	6	55	1	0,5	0,5	0,5
6	9.65 853	25	9.70 904	31	0.29 096	9.94 949	7	54	2	1,1	1,0	1,0
7	9.65 878	25	9.70 935	31	0.29 065	9.94 943	6	53	3	1,6	1,6	1,5
8	9.65 902	24	9.70 966	31	0.29 034	9.94 936	7	52	4	2,1	2,1	2,0
9	9.65 927	25	9.70 997	31	0.29 003	9.94 930	6	51	5	2,7	2,6	2,5
10	9.65 952	25	9.71 028	31	0.28 972	9.94 923	7	50	6	3,2	3,1	3,0
11	9.65 976	24	9.71 059	31	0.28 941	9.94 917	6	49	7	3,7	3,6	3,5
12	9.66 001	25	9.71 090	31	0.28 910	9.94 911	6	48	8	4,3	4,1	4,0
13	9.66 025	24	9.71 121	31	0.28 879	9.94 904	7	47	9	4,8	4,6	4,5
14	9.66 050	25	9.71 153	32	0.28 847	9.94 898	6	46			
15	9.66 075	25	9.71 184	31	0.28 816	9.94 891	7	45	10	5,3	5,2	5,0
16	9.66 099	24	9.71 215	31	0.28 785	9.94 885	6	44	20	10,7	10,3	10,0
17	9.66 124	25	9.71 246	31	0.28 754	9.94 878	7	43	30	16,0	15,5	15,0
18	9.66 148	24	9.71 277	31	0.28 723	9.94 871	7	42	40	21,3	20,7	20,0
19	9.66 173	25	9.71 308	31	0.28 692	9.94 865	6	41	50	26,7	25,8	25,0
20	9.66 197	24	9.71 339	31	0.28 661	9.94 858	7	40			
21	9.66 221	24	9.71 370	31	0.28 630	9.94 852	6	39			
22	9.66 246	25	9.71 401	30	0.28 599	9.94 845	7	38			
23	9.66 270	24	9.71 431	31	0.28 569	9.94 839	6	37		**25**	**24**	**23**
24	9.66 295	25	9.71 462	31	0.28 538	9.94 832	7	36			
25	9.66 319	24	9.71 493	31	0.28 507	9.94 826	6	35	1	0,4	0,4	0,4
26	9.66 343	25	9.71 524	31	0.28 476	9.94 819	7	34	2	0,8	0,8	0,8
27	9.66 368	24	9.71 555	31	0.28 445	9.94 813	7	33	3	1,2	1,2	1,2
28	9.66 392	24	9.71 586	31	0.28 414	9.94 806	6	32	4	1,7	1,6	1,5
29	9.66 416	25	9.71 617	31	0.28 383	9.94 799	7	31	5	2,1	2,0	1,9
30	9.66 441	24	9.71 648	31	0.28 352	9.94 793	7	30	6	2,5	2,4	2,3
31	9.66 465	24	9.71 679	30	0.28 321	9.94 786	6	29	7	2,9	2,8	2,7
32	9.66 489	24	9.71 709	31	0.28 291	9.94 780	7	28	8	3,3	3,2	3,1
33	9.66 513	24	9.71 740	31	0.28 260	9.94 773	6	27	9	3,8	3,6	3,4
34	9.66 537	25	9.71 771	31	0.28 229	9.94 767	7	26			
35	9.66 562	24	9.71 802	31	0.28 198	9.94 760	7	25	10	4,2	4,0	3,8
36	9.66 586	24	9.71 833	30	0.28 167	9.94 753	6	24	20	8,3	8,0	7,7
37	9.66 610	24	9.71 863	31	0.28 137	9.94 747	7	23	30	12,5	12,0	11,5
38	9.66 634	24	9.71 894	31	0.28 106	9.94 740	6	22	40	16,7	16,0	15,3
39	9.66 658	24	9.71 925	30	0.28 075	9.94 734	7	21	50	20,8	20,0	19,2
40	9.66 682	24	9.71 955	31	0.28 045	9.94 727	7	20			
41	9.66 706	25	9.71 986	31	0.28 014	9.94 720	6	19			
42	9.66 731	24	9.72 017	31	0.27 983	9.94 714	7	18			
43	9.66 755	24	9.72 048	30	0.27 952	9.94 707	7	17			7	6
44	9.66 779	24	9.72 078	31	0.27 922	9.94 700	6	16			
45	9.66 803	24	9.72 109	31	0.27 891	9.94 694	7	15	1	0,1	0,1	
46	9.66 827	24	9.72 140	30	0.27 860	9.94 687	7	14	2	0,2	0,2	
47	9.66 851	24	9.72 170	31	0.27 830	9.94 680	6	13	3	0,4	0,3	
48	9.66 875	24	9.72 201	30	0.27 799	9.94 674	7	12	4	0,5	0,4	
49	9.66 899	23	9.72 231	31	0.27 769	9.94 667	7	11	5	0,6	0,5	
50	9.66 922	24	9.72 262	31	0.27 738	9.94 660	6	10	6	0,7	0,6	
51	9.66 946	24	9.72 293	30	0.27 707	9.94 654	7	9	7	0,8	0,7	
52	9.66 970	24	9.72 323	31	0.27 677	9.94 647	7	8	8	0,9	0,8	
53	9.66 994	24	9.72 354	30	0.27 646	9.94 640	6	7	9	1,0	0,9	
54	9.67 018	24	9.72 384	31	0.27 616	9.94 634	7	6			
55	9.67 042	24	9.72 415	30	0.27 585	9.94 627	7	5	10	1,2	1,0	
56	9.67 066	24	9.72 445	31	0.27 555	9.94 620	6	4	20	2,3	2,0	
57	9.67 090	23	9.72 476	30	0.27 524	9.94 614	7	3	30	3,5	3,0	
58	9.67 113	24	9.72 506	31	0.27 494	9.94 607	7	2	40	4,7	4,0	
59	9.67 137	24	9.72 537	30	0.27 463	9.94 600	7	1	50	5,8	5,0	
60	9.67 161		9.72 567		0.27 433	9.94 593		0			
′	L. Cos.	d.	L.Cotg.	c.d.	L. Tang.	L. Sin.	d.	′	P. P.		

28°

′	L. Sin.	d.	L. Tang.	c.d.	L. Cotg.	L. Cos.	d.	′		P. P.	
0	9.67 161		9.72 567		0.27 433	9.94 598		60			
1	9.67 185	24	9.72 598	31	0.27 402	9.94 587	6	59			
2	9.67 208	23	9.72 628	30	0.27 372	9.94 580	7	58			
3	9.67 232	24	9.72 659	31	0.27 341	9.94 573	7	57			
4	9.67 256	24	9.72 689	30	0.27 311	9.94 567	6	56		**31**	**30**	**29**
5	9.67 280	24	9.72 720	31	0.27 280	9.94 560	7	55	1	0,5 0,5 0,5
6	9.67 303	23	9.72 750	30	0.27 250	9.94 553	7	54	2	1,0 1,0 1,0
7	9.67 327	24	9.72 780	30	0.27 220	9.94 546	7	53	3	1,6 1,5 1,4
8	9.67 350	23	9.72 811	31	0.27 189	9.94 539	6	52	4	2,1 2,0 1,9
9	9.67 374	24	9.72 841	30	0.27 159	9.94 533	7	51			
10	9.67 398	24	9.72 872	31	0.27 128	9.94 526	7	50	5	2,6 2,5 2,4
11	9.67 421	23	9.72 902	30	0.27 098	9.94 519	7	49	6	3,1 3,0 2,9
12	9.67 445	24	9.72 932	30	0.27 068	9.94 518	6	48	7	3,6 3,5 3,4
13	9.67 468	23	9.72 963	31	0.27 037	9.94 506	7	47	8	4,1 4,0 3,9
14	9.67 492	24	9.72 993	30	0.27 007	9.94 499	7	46	9	4,6 4,5 4,4
15	9.67 515	23	9.73 023	30	0.26 977	9.94 492	7	45	10	5,2 5,0 4,8
16	9.67 539	24	9.73 054	31	0.26 946	9.94 485	6	44	20	10,3 10,0 9,7
17	9.67 562	23	9.73 084	30	0.26 916	9.94 479	7	43	30	15,5 15,0 14,5
18	9.67 586	24	9.73 114	30	0.26 886	9.94 472	7	42	40	20,7 20,0 19,3
19	9.67 609	23	9.73 144	31	0.26 856	9.94 465	7	41	50	25,8 25,0 24,2
20	9.67 633	24	9.73 175	30	0.26 825	9.94 458	7	40			
21	9.67 656	23	9.73 205	30	0.26 795	9.94 451	6	39			
22	9.67 680	24	9.73 235	30	0.26 765	9.94 445	7	38			
23	9.67 703	23	9.73 265	30	0.26 735	9.94 438	7	37		**24**	**23**	**22**
24	9.67 726	24	9.73 295	31	0.26 705	9.94 431	7	36			
25	9.67 750	23	9.73 326	30	0.26 674	9.94 424	7	35	1	0,4 0,4 0,4
26	9.67 773	23	9.73 356	30	0.26 644	9.94 417	7	34	2	0,8 0,8 0,7
27	9.67 796	24	9.73 386	30	0.26 614	9.94 410	6	33	3	1,2 1,2 1,1
28	9.67 820	23	9.73 416	30	0.26 584	9.94 404	7	32	4	1,6 1,5 1,5
29	9.67 843	23	9.73 446	30	0.26 554	9.94 397	7	31			
30	9.67 866	24	9.73 476	31	0.26 524	9.94 390	7	30	5	2,0 1,9 1,8
31	9.67 890	23	9.73 507	30	0.26 493	9.94 383	7	29	6	2,4 2,3 2,2
32	9.67 913	23	9.73 537	30	0.26 463	9.94 376	7	28	7	2,8 2,7 2,6
33	9.67 936	23	9.73 567	30	0.26 433	9.94 369	7	27	8	3,2 3,1 2,9
34	9.67 959	23	9.73 597	30	0.26 403	9.94 362	7	26	9	3,6 3,4 3,3
35	9.67 982	24	9.73 627	30	0.26 373	9.94 355	6	25	10	4,0 3,8 3,7
36	9.68 006	23	9.73 657	30	0.26 343	9.94 349	7	24	20	8,0 7,7 7,3
37	9.68 029	23	9.73 687	30	0.26 313	9.94 342	7	23	30	12,0 11,5 11,0
38	9.68 052	23	9.73 717	30	0.26 283	9.94 335	7	22	40	16,0 15,3 14,7
39	9.68 075	23	9.73 747	30	0.26 253	9.94 328	7	21	50	20,0 19,2 18,3
40	9.68 098	23	9.73 777	30	0.26 223	9.94 321	7	20			
41	9.68 121	23	9.73 807	30	0.26 193	9.94 314	7	19			
42	9.68 144	23	9.73 837	30	0.26 163	9.94 307	7	18			
43	9.68 167	23	9.73 867	30	0.26 133	9.94 300	7	17		**7**	**6**
44	9.68 190	23	9.73 897	30	0.26 103	9.94 293	7	16			
45	9.68 213	24	9.73 927	30	0.26 073	9.94 286	7	15	1	0,1 0,1
46	9.68 237	23	9.73 957	30	0.26 043	9.94 279	6	14	2	0,2 0,2
47	9.68 260	23	9.73 987	30	0.26 013	9.94 273	7	13	3	0,4 0,3
48	9.68 283	22	9.74 017	30	0.25 983	9.94 266	7	12	4	0,5 0,4
49	9.68 305	23	9.74 047	30	0.25 953	9.94 259	7	11			
50	9.68 328	23	9.74 077	30	0.25 923	9.94 252	7	10	5	0,6 0,5
51	9.68 351	23	9.74 107	30	0.25 893	9.94 245	7	9	6	0,7 0,6
52	9.68 374	23	9.74 137	29	0.25 863	9.94 238	7	8	7	0,8 0,7
53	9.68 397	23	9.74 166	30	0.25 834	9.94 231	7	7	8	0,9 0,8
54	9.68 420	23	9.74 196	30	0.25 804	9.94 224	7	6	9	1,0 0,9
55	9.68 443	23	9.74 226	30	0.25 774	9.94 217	7	5			
56	9.68 466	23	9.74 256	30	0.25 744	9.94 210	7	4	10	1,2 1,0
57	9.68 489	23	9.74 286	30	0.25 714	9.94 203	7	3	20	2,3 2,0
58	9.68 512	22	9.74 316	29	0.25 684	9.94 196	7	2	30	3,5 3,0
59	9.68 534	23	9.74 345	30	0.25 655	9.94 189	7	1	40	4,7 4,0
60	9.68 557		9.74 375		0.25 625	9.94 182		0	50	5,8 5,0

| | L. Cos. | d. | L. Cotg. | c.d. | L. Tang. | L. Sin. | d. | ′ | | P. P. | |

61°

29°

'	L. Sin.	d.	L. Tang.	c.d.	L. Cotg.	L. Cos.	d.		P. P.		
0	9.68 557	28	9.74 375	30	0.25 625	9.94 182	7	**60**			
1	9.68 580	28	9.74 405	30	0.25 595	9.94 175	7	59			
2	9.68 608	22	9.74 435	30	0.25 565	9.94 168	7	58			
3	9.68 625	23	9.74 465	29	0.25 535	9.94 161	7	57			
4	9.68 648	28	9.74 494	30	0.25 506	9.94 154	7	56			
5	9.68 671	28	9.74 524	30	0.25 476	9.94 147	7	55			
6	9.68 694	22	9.74 554	29	0.25 446	9.94 140	7	54			
7	9.68 716	28	9.74 583	30	0.25 417	9.94 133	7	53			
8	9.68 739	28	9.74 613	30	0.25 387	9.94 126	7	52			
9	9.68 762	22	9.74 643	30	0.25 357	9.94 119	7	51			
10	9.68 784	28	9.74 673	29	0.25 327	9.94 112	7	**50**	**30**	**20**	**23**
11	9.68 807	22	9.74 702	30	0.25 298	9.94 105	7	49			
12	9.68 829	28	9.74 732	30	0.25 268	9.94 098	8	48	1 0,5	0,5	0,4
13	9.68 852	23	9.74 762	29	0.25 238	9.94 090	7	47	2 1,0	1,0	0,8
14	9.68 875	22	9.74 791	30	0.25 209	9.94 083	7	46	3 1,5	1,4	1,2
									4 2,0	1,9	1,5
15	9.68 897	28	9.74 821	30	0.25 179	9.94 076	7	45			
16	9.68 920	22	9.74 851	29	0.25 149	9.94 069	7	44	5 2,5	2,4	1,9
17	9.68 942	23	9.74 880	30	0.25 120	9.94 062	7	43	6 3,0	2,9	2,3
18	9.68 965	22	9.74 910	29	0.25 090	9.94 055	7	42	7 3,5	3,4	2,7
19	9.68 987	28	9.74 939	30	0.25 061	9.94 048	7	41	8 4,0	3,9	3,1
									9 4,5	4,4	3,4
20	9.69 010	22	9.74 969	29	0.25 031	9.94 041	7	**40**			
21	9.69 032	23	9.74 998	30	0.25 002	9.94 034	7	39	10 5,0	4,8	3,8
22	9.69 055	22	9.75 028	30	0.24 972	9.94 027	7	38	20 10,0	9,7	7,7
23	9.69 077	28	9.75 058	29	0.24 942	9.94 020	8	37	30 15,0	14,5	11,5
24	9.69 100	22	9.75 087	30	0.24 913	9.94 012	7	36	40 20,0	19,3	15,3
25	9.69 122	22	9.75 117	29	0.24 883	9.94 005	7	35	50 25,0	24,2	19,2
26	9.69 144	23	9.75 146	30	0.24 854	9.93 998	7	34			
27	9.69 167	22	9.75 176	29	0.24 824	9.93 991	7	33			
28	9.69 189	23	9.75 205	30	0.24 795	9.93 984	7	32			
29	9.69 212	22	9.75 235	29	0.24 765	9.93 977	7	31			
30	9.69 234	22	9.75 264	30	0.24 736	9.93 970	7	**30**			
31	9.69 256	23	9.75 294	29	0.24 706	9.93 963	8	29			
32	9.69 279	22	9.75 323	30	0.24 677	9.93 955	7	28			
33	9.69 301	22	9.75 353	30	0.24 647	9.93 948	7	27			
34	9.69 323	22	9.75 382	29	0.24 618	9.93 941	7	26			
35	9.69 345	23	9.75 411	30	0.24 589	9.93 934	7	25	**22**	**8**	**7**
36	9.69 368	22	9.75 441	29	0.24 559	9.93 927	7	24			
37	9.69 390	22	9.75 470	30	0.24 530	9.93 920	8	23	1 0,4	0,1	0,1
38	9.69 412	22	9.75 500	29	0.24 500	9.93 912	7	22	2 0,7	0,3	0,2
39	9.69 434	22	9.75 529	29	0.24 471	9.93 905	7	21	3 1,1	0,4	0,4
									4 1,5	0,5	0,5
40	9.69 456	23	9.75 558	30	0.24 442	9.93 898	7	**20**			
41	9.69 479	22	9.75 588	29	0.24 412	9.93 891	7	19	5 1,8	0,7	0,6
42	9.69 501	22	9.75 617	30	0.24 383	9.93 884	8	18	6 2,2	0,8	0,7
43	9.69 523	22	9.75 647	29	0.24 353	9.93 876	7	17	7 2,6	0,9	0,8
44	9.69 545	22	9.75 676	29	0.24 324	9.93 869	7	16	8 2,9	1,1	0,9
									9 3,3	1,2	1,0
45	9.69 567	22	9.75 705	30	0.24 295	9.93 862	7	15			
46	9.69 589	22	9.75 735	29	0.24 265	9.93 855	8	14	10 3,7	1,3	1,2
47	9.69 611	22	9.75 764	29	0.24 236	9.93 847	7	13	20 7,3	2,7	2,3
48	9.69 633	22	9.75 793	30	0.24 207	9.93 840	7	12	30 11,0	4,0	3,5
49	9.69 655	22	9.75 822	30	0.24 178	9.93 833	7	11	40 14,7	5,3	4,7
50	9.69 677	22	9.75 852	29	0.24 148	9.93 826	7	**10**	50 18,3	6,7	5,8
51	9.69 699	22	9.75 881	29	0.24 119	9.93 819	8	9			
52	9.69 721	22	9.75 910	29	0.24 090	9.93 811	7	8			
53	9.69 743	22	9.75 939	30	0.24 061	9.93 804	7	7			
54	9.69 765	22	9.75 969	29	0.24 031	9.93 797	8	6			
55	9.69 787	22	9.75 998	29	0.24 002	9.93 789	7	5			
56	9.69 809	22	9.76 027	29	0.23 973	9.93 782	7	4			
57	9.69 831	22	9.76 056	30	0.23 944	9.93 775	7	3			
58	9.69 853	22	9.76 086	29	0.23 914	9.93 768	8	2			
59	9.69 875	22	9.76 115	29	0.23 885	9.93 760	7	1			
60	9.69 897		9.76 144		0.23 856	9.93 753		**0**			
	L. Cos.	d.	L. Cotg.	c.d.	L. Tang.	L. Sin.	d.	'	P. P.		

60°

30°

′	L. Sin.	d.	L. Tang.	c.d.	L. Cotg.	L. Cos.	d.		P. P.		
0	9.69 697		9.76 144		0.23 856	9.93 753		60			
1	9.69 919	22	9.76 173	29	0.23 827	9.93 746	7	59			
2	9.69 941	22	9.76 202	29	0.23 798	9.93 738	8	58			
3	9.69 963	22	9.76 231	29	0.23 769	9.93 731	7	57			
4	9.69 984	21	9.76 261	30	0.23 789	9.93 724	7	56		30	29	28
5	9.70 006	22	9.76 290	29	0.23 710	9.93 717	7	55	1	0,5	0,5	0,5
6	9.70 028	22	9.76 319	29	0.23 681	9.93 709	8	54	2	1,0	1,0	0,9
7	9.70 050	22	9.76 348	29	0.23 652	9.93 702	7	53	3	1,5	1,4	1,4
8	9.70 072	21	9.76 377	29	0.23 623	9.93 695	7	52	4	2,0	1,9	1,9
9	9.70 093	22	9.76 406	29	0.23 594	9.93 687	8	51				
10	9.70 115	22	9.76 435	29	0.23 565	9.93 680	7	50	5	2,5	2,4	2,3
11	9.70 137	22	9.76 464	29	0.23 536	9.93 673	7	49	6	3,0	2,9	2,8
12	9.70 159	21	9.76 493	29	0.23 507	9.93 665	8	48	7	3,5	3,4	3,3
13	9.70 180	22	9.76 522	29	0.23 478	9.93 658	7	47	8	4,0	3,9	3,7
14	9.70 202	22	9.76 551	29	0.23 449	9.93 650	7	46	9	4,5	4,4	4,2
15	9.70 224	21	9.76 580	29	0.23 420	9.93 643	7	45				
16	9.70 245	22	9.76 609	30	0.23 391	9.93 636	8	44	10	5,0	4,8	4,7
17	9.70 267	21	9.76 639	29	0.23 361	9.93 628	7	43	20	10,0	9,7	9,3
18	9.70 288	22	9.76 668	29	0.23 332	9.93 621	7	42	30	15,0	14,5	14,0
19	9.70 310	22	9.76 697	28	0.23 303	9.93 614	8	41	40	20,0	19,3	18,7
20	9.70 332	21	9.76 725	29	0.23 275	9.93 606	7	40	50	25,0	24,2	23,3
21	9.70 353	22	9.76 754	29	0.23 246	9.93 599	8	39				
22	9.70 375	21	9.76 783	29	0.23 217	9.93 591	7	38				
23	9.70 396	22	9.76 812	29	0.23 188	9.93 584	7	37			22	21
24	9.70 418	21	9.76 841	29	0.23 159	9.93 577	8	36				
25	9.70 439	22	9.76 870	29	0.23 130	9.93 569	7	35	1		0,4	0,4
26	9.70 461	21	9.76 899	29	0.23 101	9.93 562	8	34	2		0,7	0,7
27	9.70 482	22	9.76 928	29	0.23 072	9.93 554	7	33	3		1,1	1,0
28	9.70 504	21	9.76 957	29	0.23 043	9.93 547	8	32	4		1,5	1,4
29	9.70 525	22	9.76 986	29	0.23 014	9.93 539	7	31				
30	9.70 547	21	9.77 015	29	0.22 985	9.93 532	7	30	5		1,8	1,8
31	9.70 568	22	9.77 044	29	0.22 956	9.93 525	8	29	6		2,2	2,1
32	9.70 590	21	9.77 073	28	0.22 927	9.93 517	7	28	7		2,6	2,4
33	9.70 611	22	9.77 101	29	0.22 899	9.93 510	8	27	8		2,9	2,8
34	9.70 633	21	9.77 130	29	0.22 870	9.93 502	7	26	9		3,3	3,2
35	9.70 654	21	9.77 159	29	0.22 841	9.93 495	8	25				
36	9.70 675	22	9.77 188	29	0.22 812	9.93 487	7	24	10		3,7	3,5
37	9.70 697	21	9.77 217	29	0.22 783	9.93 480	8	23	20		7,3	7,0
38	9.70 718	22	9.77 246	28	0.22 754	9.93 472	7	22	30		11,0	10,5
39	9.70 739	22	9.77 274	29	0.22 726	9.93 465	8	21	40		14,7	14,0
40	9.70 761	21	9.77 303	29	0.22 697	9.93 457	7	20	50		18,3	17,5
41	9.70 782	21	9.77 332	29	0.22 668	9.93 450	8	19				
42	9.70 803	21	9.77 361	29	0.22 639	9.93 442	7	18				
43	9.70 824	22	9.77 390	28	0.22 610	9.93 435	8	17			8	7
44	9.70 846	21	9.77 418	29	0.22 582	9.93 427	7	16				
45	9.70 867	21	9.77 447	29	0.22 553	9.93 420	8	15	1		0,1	0,1
46	9.70 888	21	9.77 476	29	0.22 524	9.93 412	7	14	2		0,3	0,2
47	9.70 909	22	9.77 505	28	0.22 495	9.93 405	8	13	3		0,4	0,4
48	9.70 931	21	9.77 533	29	0.22 467	9.93 397	7	12	4		0,5	0,5
49	9.70 952	21	9.77 562	29	0.22 438	9.93 390	8	11				
50	9.70 973	21	9.77 591	28	0.22 409	9.93 382	7	10	5		0,7	0,6
51	9.70 994	21	9.77 619	29	0.22 381	9.93 375	8	9	6		0,8	0,7
52	9.71 015	21	9.77 648	29	0.22 352	9.93 367	7	8	7		0,9	0,8
53	9.71 036	22	9.77 677	29	0.22 323	9.93 360	8	7	8		1,1	0,9
54	9.71 058	21	9.77 706	28	0.22 294	9.93 352	8	6	9		1,2	1,0
55	9.71 079	21	9.77 734	29	0.22 266	9.93 344	7	5				
56	9.71 100	21	9.77 763	28	0.22 237	9.93 337	8	4	10		1,8	1,2
57	9.71 121	21	9.77 791	29	0.22 209	9.93 329	7	3	20		2,7	2,3
58	9.71 142	21	9.77 820	29	0.22 180	9.93 322	8	2	30		4,0	3,5
59	9.71 163	21	9.77 849	28	0.22 151	9.93 314	7	1	40		5,8	4,7
60	9.71 184		9.77 877		0.22 123	9.93 307		0	50		6,7	5,8

	L. Cos.	d.	L. Cotg.	c.d.	L. Tang.	L. Sin.	d.	′	P. P.		

59°

31°

'	L. Sin.	d.	L. Tang.	c.d.	L. Cotg.	L. Cos.	d.		P. P.	
0	9.71 184	21	9.77 877	29	0.22 123	9.93 307	8	60		
1	9.71 205	21	9.77 906	29	0.22 094	9.93 299	8	59		
2	9.71 226	21	9.77 935	28	0.22 065	9.93 291	7	58		
3	9.71 247	21	9.77 963	29	0.22 037	9.93 284	8	57	**29**	**28**
4	9.71 268	21	9.77 992	28	0.22 008	9.93 276	7	56		
5	9.71 289	21	9.78 020	29	0.21 980	9.93 269	8	55	1 0,5	0,5
6	9.71 310	21	9.78 049	28	0.21 951	9.93 261	8	54	2 1,0	0,9
7	9.71 331	21	9.78 077	29	0.21 923	9.93 253	7	53	3 1,4	1,4
8	9.71 352	21	9.78 106	29	0.21 894	9.93 246	8	52	4 1,9	1,9
9	9.71 373	20	9.78 135	28	0.21 865	9.93 238	8	51		
10	9.71 393	21	9.78 163	29	0.21 837	9.93 230	7	50	5 2,4	2,3
11	9.71 414	21	9.78 192	28	0.21 808	9.93 223	8	49	6 2,9	2,8
12	9.71 435	21	9.78 220	29	0.21 780	9.93 215	8	48	7 3,4	3,3
13	9.71 456	21	9.78 249	28	0.21 751	9.93 207	7	47	8 3,9	3,7
14	9.71 477	21	9.78 277	29	0.21 723	9.93 200	8	46	9 4,4	4,2
15	9.71 498	21	9.78 306	28	0.21 694	9.93 192	8	45	10 4,8	4,7
16	9.71 519	20	9.78 334	29	0.21 666	9.93 184	7	44	20 9,7	9,3
17	9.71 539	21	9.78 363	28	0.21 637	9.93 177	8	43	30 14,5	14,0
18	9.71 560	21	9.78 391	28	0.21 609	9.93 169	8	42	40 19,3	18,7
19	9.71 581	21	9.78 419	29	0.21 581	9.93 161	7	41	50 24,2	23,3
20	9.71 602	20	9.78 448	28	0.21 552	9.93 154	8	40		
21	9.71 622	21	9.78 476	29	0.21 524	9.93 146	8	39		
22	9.71 643	21	9.78 505	28	0.21 495	9.93 138	7	38		
23	9.71 664	21	9.78 533	29	0.21 467	9.93 131	8	37	**21**	**20**
24	9.71 685	20	9.78 562	28	0.21 438	9.93 123	8	36		
25	9.71 705	21	9.78 590	28	0.21 410	9.93 115	7	35	1 0,4	0,3
26	9.71 726	21	9.78 618	29	0.21 382	9.93 108	8	34	2 0,7	0,7
27	9.71 747	20	9.78 647	28	0.21 353	9.93 100	8	33	3 1,0	1,0
28	9.71 767	21	9.78 675	29	0.21 325	9.93 092	8	32	4 1,4	1,3
29	9.71 788	21	9.78 704	28	0.21 296	9.93 084	7	31		
30	9.71 809	20	9.78 732	28	0.21 268	9.93 077	8	30	5 1,8	1,7
31	9.71 829	21	9.78 760	29	0.21 240	9.93 069	8	29	6 2,1	2,0
32	9.71 850	20	9.78 789	28	0.21 211	9.93 061	8	28	7 2,4	2,3
33	9.71 870	21	9.78 817	28	0.21 183	9.93 053	7	27	8 2,8	2,7
34	9.71 891	20	9.78 845	29	0.21 155	9.93 046	8	26	9 3,2	3,0
35	9.71 911	21	9.78 874	28	0.21 126	9.93 038	8	25	10 3,5	3,3
36	9.71 932	20	9.78 902	28	0.21 098	9.93 030	8	24	20 7,0	6,7
37	9.71 952	21	9.78 930	29	0.21 070	9.93 022	8	23	30 10,5	10,0
38	9.71 973	20	9.78 959	28	0.21 041	9.93 014	7	22	40 14,0	13,3
39	9.71 994	20	9.78 987	28	0.21 013	9.93 007	8	21	50 17,5	16,7
40	9.72 014	20	9.79 015	29	0.20 985	9.92 999	8	20		
41	9.72 034	21	9.79 043	29	0.20 957	9.92 991	8	19		
42	9.72 055	20	9.79 072	28	0.20 928	9.92 983	7	18	**8**	**7**
43	9.72 075	21	9.79 100	28	0.20 900	9.92 976	8	17		
44	9.72 096	20	9.79 128	28	0.20 872	9.92 968	8	16	1 0,1	0,1
45	9.72 116	21	9.79 156	29	0.20 844	9.92 960	8	15	2 0,3	0,2
46	9.72 137	20	9.79 185	28	0.20 815	9.92 952	8	14	3 0,4	0,4
47	9.72 157	20	9.79 213	28	0.20 787	9.92 944	8	13	4 0,5	0,5
48	9.72 177	21	9.79 241	28	0.20 759	9.92 936	7	12		
49	9.72 198	20	9.79 269	28	0.20 731	9.92 929	8	11	5 0,7	0,6
50	9.72 218	20	9.79 297	29	0.20 703	9.92 921	8	10	6 0,8	0,7
51	9.72 238	21	9.79 326	28	0.20 674	9.92 913	8	9	7 0,9	0,8
52	9.72 259	20	9.79 354	28	0.20 646	9.92 905	8	8	8 1,1	0,9
53	9.72 279	20	9.79 382	28	0.20 618	9.92 897	8	7	9 1,2	1,0
54	9.72 299	21	9.79 410	28	0.20 590	9.92 889	8	6		
55	9.72 320	20	9.79 438	28	0.20 562	9.92 881	7	5	10 1,3	1,2
56	9.72 340	20	9.79 466	29	0.20 534	9.92 874	8	4	20 2,7	2,3
57	9.72 360	21	9.79 495	28	0.20 505	9.92 866	8	3	30 4,0	3,5
58	9.72 381	20	9.79 523	28	0.20 477	9.92 858	8	2	40 5,3	4,7
59	9.72 401	20	9.79 551	28	0.20 449	9.92 850	8	1	50 6,7	5,8
60	9.72 421		9.79 579		0.20 421	9.92 842		0		

| | L. Cos. | d. | L. Cotg. | c.d. | L. Tang. | L. Sin. | d. | ' | P. P. | |

58°

'	L. Sin.	d.	L. Tang.	c.d.	L. Cotg.	L. Cos.	d.		P. P.		
0	9.72 421	20	9.79 579	28	0.20 421	9.92 842	8	60			
1	9.72 441	20	9.79 607	28	0.20 393	9.92 834	8	59			
2	9.72 461	21	9.79 635	28	0.20 365	9.92 826	8	58			
3	9.72 482	20	9.79 663	28	0.20 337	9.92 818	8	57		**29**	**28**	**27**
4	9.72 502	20	9.79 691	28	0.20 309	9.92 810	7	56	1	0,5	0,5	0,4
5	9.72 522	20	9.79 719	28	0.20 281	9.92 803	8	55	2	1,0	0,9	0,9
6	9.72 542	20	9.79 747	29	0.20 253	9.92 795	8	54	3	1,4	1,4	1,4
7	9.72 562	20	9.79 776	28	0.20 224	9.92 787	8	53	4	1,9	1,9	1,8
8	9.72 582	20	9.79 804	28	0.20 196	9.92 779	8	52				
9	9.72 602	20	9.79 832	28	0.20 168	9.92 771	8	51	5	2,4	2,3	2,2
10	9.72 622	21	9.79 860	28	0.20 140	9.92 763	8	50	6	2,9	2,8	2,7
11	9.72 643	20	9.79 888	28	0.20 112	9.92 755	8	49	7	3,4	3,3	3,2
12	9.72 663	20	9.79 916	28	0.20 084	9.92 747	8	48	8	3,9	3,7	3,6
13	9.72 683	20	9.79 944	28	0.20 056	9.92 739	8	47	9	4,4	4,2	4,0
14	9.72 703	20	9.79 972	28	0.20 028	9.92 731	8	46				
15	9.72 723	20	9.80 000	28	0.20 000	9.92 723	8	45	10	4,8	4,7	4,5
16	9.72 743	20	9.80 028	28	0.19 972	9.92 715	8	44	20	9,7	9,3	9,0
17	9.72 763	20	9.80 056	28	0.19 944	9.92 707	8	43	30	14,5	14,0	13,5
18	9.72 783	20	9.80 084	28	0.19 916	9.92 699	8	42	40	19,3	18,7	18,0
19	9.72 803	20	9.80 112	28	0.19 888	9.92 691	8	41	50	24,2	23,3	22,5
20	9.72 823	20	9.80 140	28	0.19 860	9.92 683	8	40				
21	9.72 843	20	9.80 168	27	0.19 832	9.92 675	8	39				
22	9.72 863	20	9.80 195	28	0.19 805	9.92 667	8	38				
23	9.72 883	19	9.80 223	28	0.19 777	9.92 659	8	37		**21**	**20**	**19**
24	9.72 902	20	9.80 251	28	0.19 749	9.92 651	8	36	1	0,4	0,3	0,3
25	9.72 922	20	9.80 279	28	0.19 721	9.92 643	8	35	2	0,7	0,7	0,6
26	9.72 942	20	9.80 307	28	0.19 693	9.92 635	8	34	3	1,0	1,0	1,0
27	9.72 962	20	9.80 335	28	0.19 665	9.92 627	8	33	4	1,4	1,3	1,3
28	9.72 982	20	9.80 363	28	0.19 637	9.92 619	8	32				
29	9.73 002	20	9.80 391	28	0.19 609	9.92 611	8	31	5	1,8	1,7	1,6
30	9.73 022	19	9.80 419	28	0.19 581	9.92 603	8	30	6	2,1	2,0	1,9
31	9.73 041	20	9.80 447	27	0.19 553	9.92 595	8	29	7	2,4	2,3	2,2
32	9.73 061	20	9.80 474	28	0.19 526	9.92 587	8	28	8	2,8	2,7	2,5
33	9.73 081	20	9.80 502	28	0.19 498	9.92 579	8	27	9	3,2	3,0	2,8
34	9.73 101	20	9.80 530	28	0.19 470	9.92 571	8	26				
35	9.73 121	19	9.80 558	28	0.19 442	9.92 563	8	25	10	3,5	3,3	3,2
36	9.73 140	20	9.80 586	28	0.19 414	9.92 555	9	24	20	7,0	6,7	6,3
37	9.73 160	20	9.80 614	28	0.19 386	9.92 546	8	23	30	10,5	10,0	9,5
38	9.73 180	20	9.80 642	27	0.19 358	9.92 538	8	22	40	14,0	13,3	12,7
39	9.73 200	19	9.80 669	28	0.19 331	9.92 530	8	21	50	17,5	16,7	15,8
40	9.73 219	20	9.80 697	28	0.19 303	9.92 522	8	20				
41	9.73 239	20	9.80 725	28	0.19 275	9.92 514	8	19				
42	9.73 259	19	9.80 753	28	0.19 247	9.92 506	8	18		**9**	**8**	**7**
43	9.73 278	20	9.80 781	27	0.19 219	9.92 498	8	17	1	0,2	0,1	0,1
44	9.73 298	20	9.80 808	28	0.19 192	9.92 490	8	16	2	0,3	0,3	0,2
45	9.73 318	19	9.80 836	28	0.19 164	9.92 482	9	15	3	0,4	0,4	0,4
46	9.73 337	20	9.80 864	28	0.19 136	9.92 473	8	14	4	0,6	0,5	0,5
47	9.73 357	20	9.80 892	27	0.19 108	9.92 465	8	13				
48	9.73 377	19	9.80 919	28	0.19 081	9.92 457	8	12	5	0,8	0,7	0,6
49	9.73 396	20	9.80 947	28	0.19 053	9.92 449	8	11	6	0,9	0,8	0,7
50	9.73 416	19	9.80 975	28	0.19 025	9.92 441	8	10	7	1,0	0,9	0,8
51	9.73 435	20	9.81 003	27	0.18 997	9.92 433	8	9	8	1,2	1,1	0,9
52	9.73 455	19	9.81 030	28	0.18 970	9.92 425	9	8	9	1,4	1,2	1,0
53	9.73 474	20	9.81 058	28	0.18 942	9.92 416	8	7				
54	9.73 494	19	9.81 086	27	0.18 914	9.92 408	8	6	10	1,5	1,3	1,2
55	9.73 513	20	9.81 118	23	0.18 887	9.92 400	8	5	20	3,0	2,7	2,3
56	9.73 533	19	9.81 141	28	0.18 859	9.92 392	8	4	30	4,5	4,0	3,5
57	9.73 552	20	9.81 169	27	0.18 831	9.92 384	8	3	40	6,0	5,3	4,7
58	9.73 572	19	9.81 196	28	0.18 804	9.92 376	9	2	50	7,5	6,7	5,8
59	9.73 591	20	9.81 224	28	0.18 776	9.92 367	8	1				
60	9.73 611		9.81 252		0.18 748	9.92 359		0				
	L. Cos.	d.	L. Cotg.	c.d.	L. Tang.	L. Sin.	d.	'		P. P.		

'	L. Sin.	d.	L. Tang.	c.d.	L. Cotg.	L. Cos.	d.	
0	9.78 611		9.81 252		0.18 748	9.92 359		60
1	9.78 630	19	9.81 279	27	0.18 721	9.92 351	8	59
2	9.78 650	20	9.81 307	28	0.18 693	9.92 348	8	58
3	9.78 669	19	9.81 335	28	0.18 665	9.92 335	8	57
4	9.78 689	20	9.81 362	27	0.18 638	9.92 326	9	56
		19		28			8	
5	9.78 708		9.81 390		0.18 610	9.92 318		55
6	9.78 727	19	9.81 418	28	0.18 582	9.92 310	8	54
7	9.78 747	20	9.81 445	27	0.18 555	9.92 302	8	53
8	9.78 766	19	9.81 473	23	0.18 527	9.92 293	9	52
9	9.78 785	19	9.81 500	27	0.18 500	9.92 285	8	51
		20					8	
10	9.78 805		9.81 528		0.18 472	9.92 277		50
11	9.78 824	19	9.81 556	28	0.18 444	9.92 269	8	49
12	9.78 843	19	9.81 583	27	0.18 417	9.92 260	9	48
13	9.78 863	20	9.81 611	28	0.18 389	9.92 252	8	47
14	9.78 882	19	9.81 638	27	0.18 362	9.92 244	8	46
		19		28				
15	9.78 901		9.81 666		0.18 334	9.92 235		45
16	9.78 921	20	9.81 693	27	0.18 307	9.92 227	8	44
17	9.78 940	19	9.81 721	28	0.18 279	9.92 219	8	43
18	9.78 959	19	9.81 748	27	0.18 252	9.92 211	8	42
19	9.78 978	19	9.81 776	28	0.18 224	9.92 202	9	41
		19		27			8	
20	9.78 997		9.81 803		0.18 197	9.92 194		40
21	9.74 017	20	9.81 831	28	0.18 169	9.92 186	8	39
22	9.74 036	19	9.81 858	27	0.18 142	9.92 177	9	38
23	9.74 055	19	9.81 886	28	0.18 114	9.92 169	8	37
24	9.74 074	19	9.81 913	27	0.18 087	9.92 161	8	36
		19		28			9	
25	9.74 093		9.81 941		0.18 059	9.92 152		35
26	9.74 113	20	9.81 968	27	0.18 032	9.92 144	8	34
27	9.74 132	19	9.81 996	28	0.18 004	9.92 136	8	33
28	9.74 151	19	9.82 023	27	0.17 977	9.92 127	8	32
29	9.74 170	19	9.82 051	28	0.17 949	9.92 119	9	31
		19		27			8	
30	9.74 189		9.82 078		0.17 922	9.92 111		30
31	9.74 208	19	9.82 106	28	0.17 894	9.92 102	9	29
32	9.74 227	19	9.82 133	27	0.17 867	9.92 094	8	28
33	9.74 246	19	9.82 161	28	0.17 839	9.92 086	8	27
34	9.74 265	19	9.82 188	27	0.17 812	9.92 077	9	26
		19		27			8	
35	9.74 284		9.82 215		0.17 785	9.92 069		25
36	9.74 303	19	9.82 243	28	0.17 757	9.92 060	9	24
37	9.74 322	19	9.82 270	27	0.17 730	9.92 052	8	23
38	9.74 341	19	9.82 298	28	0.17 702	9.92 044	8	22
39	9.74 360	19	9.82 325	27	0.17 675	9.92 035	9	21
		19		27			8	
40	9.74 379		9.82 352		0.17 648	9.92 027		20
41	9.74 398	19	9.82 380	23	0.17 620	9.92 018	9	19
42	9.74 417	19	9.82 407	27	0.17 593	9.92 010	8	18
43	9.74 436	19	9.82 435	28	0.17 565	9.92 002	8	17
44	9.74 455	19	9.82 462	27	0.17 538	9.91 993	9	16
		19		27			8	
45	9.74 474		9.82 489		0.17 511	9.91 985		15
46	9.74 493	19	9.82 517	28	0.17 483	9.91 976	9	14
47	9.74 512	19	9.82 544	27	0.17 456	9.91 968	8	13
48	9.74 531	19	9.82 571	27	0.17 429	9.91 959	9	12
49	9.74 549	18	9.82 599	28	0.17 401	9.91 951	8	11
		19		27				
50	9.74 568		9.82 626		0.17 374	9.91 942		10
51	9.74 587	19	9.82 653	27	0.17 347	9.91 934	8	9
52	9.74 606	19	9.82 681	28	0.17 319	9.91 925	9	8
53	9.74 625	19	9.82 708	27	0.17 292	9.91 917	8	7
54	9.74 644	19	9.82 735	27	0.17 265	9.91 908	9	6
		18					9	
55	9.74 662		9.82 762		0.17 238	9.91 900		5
56	9.74 681	19	9.82 790	28	0.17 210	9.91 891	9	4
57	9.74 700	19	9.82 817	27	0.17 183	9.91 883	8	3
58	9.74 719	19	9.82 844	27	0.17 156	9.91 874	9	2
59	9.74 737	18	9.82 871	27	0.17 129	9.91 866	8	1
		19		28			9	
60	9.74 756		9.82 899		0.17 101	9.91 857		0
	L. Cos.	d.	L. Cotg.	c.d.	L. Tang.	L. Sin.	d.	'

P. P.

	28	27
1	0,5	0,4
2	0,9	0,9
3	1,4	1,4
4	1,9	1,8
5	2,3	2,2
6	2,8	2,7
7	3,3	3,2
8	3,7	3,6
9	4,2	4,0
10	4,7	4,5
20	9,3	9,0
30	14,0	13,5
40	18,7	18,0
50	23,3	22,5

	20	19	18
1	0,3	0,3	0,3
2	0,7	0,6	0,6
3	1,0	1,0	0,9
4	1,3	1,3	1,2
5	1,7	1,6	1,5
6	2,0	1,9	1,8
7	2,3	2,2	2,1
8	2,7	2,5	2,4
9	3,0	2,8	2,7
10	3,3	3,2	3,0
20	6,7	6,3	6,0
30	10,0	9,5	9,0
40	13,3	12,7	12,0
50	16,7	15,8	15,0

	9	8
1	0,2	0,1
2	0,3	0,3
3	0,4	0,4
4	0,6	0,5
5	0,8	0,7
6	0,9	0,8
7	1,0	0,9
8	1,2	1,1
9	1,4	1,2
10	1,5	1,3
20	3,0	2,7
30	4,5	4,0
40	6,0	5,3
50	7,5	6,7

P. P.

34°

′	L. Sin.	d.	L. Tang.	c.d.	L. Cotg.	L. Cos.	d.	
0	9.74 756	19	9.82 899	27	0.17 101	9.91 857	8	60
1	9.74 775	19	9.82 926	27	0.17 074	9.91 849	9	59
2	9.74 794	18	9.82 958	27	0.17 047	9.91 840	8	58
3	9.74 812	19	9.82 980	28	0.17 020	9.91 832	9	57
4	9.74 831	19	9.83 008	27	0.16 992	9.91 823	8	56
5	9.74 850	18	9.83 035	27	0.16 965	9.91 815	9	55
6	9.74 868	19	9.83 062	27	0.16 938	9.91 806	8	54
7	9.74 887	19	9.83 089	28	0.16 911	9.91 798	9	53
8	9.74 906	18	9.83 117	27	0.16 883	9.91 789	8	52
9	9.74 924	19	9.83 144	27	0.16 856	9.91 781	9	51
10	9.74 943	18	9.83 171	27	0.16 829	9.91 772	9	50
11	9.74 961	19	9.83 198	27	0.16 802	9.91 763	8	49
12	9.74 980	19	9.83 225	27	0.16 775	9.91 755	9	48
13	9.74 999	18	9.83 252	28	0.16 748	9.91 746	8	47
14	9.75 017	19	9.83 280	27	0.16 720	9.91 738	9	46
15	9.75 036	18	9.83 307	27	0.16 693	9.91 729	9	45
16	9.75 054	19	9.83 334	27	0.16 666	9.91 720	8	44
17	9.75 073	18	9.83 361	27	0.16 639	9.91 712	9	43
18	9.75 091	19	9.83 388	27	0.16 612	9.91 703	8	42
19	9.75 110	18	9.83 415	27	0.16 585	9.91 695	9	41
20	9.75 128	19	9.83 442	28	0.16 558	9.91 686	9	40
21	9.75 147	18	9.83 470	27	0.16 530	9.91 677	8	39
22	9.75 165	19	9.83 497	27	0.16 503	9.91 669	9	38
23	9.75 184	18	9.83 524	27	0.16 476	9.91 660	9	37
24	9.75 202	19	9.83 551	27	0.16 449	9.91 651	8	36
25	9.75 221	18	9.83 578	27	0.16 422	9.91 643	9	35
26	9.75 239	19	9.83 605	27	0.16 395	9.91 634	9	34
27	9.75 258	18	9.83 632	27	0.16 368	9.91 625	8	33
28	9.75 276	18	9.83 659	27	0.16 341	9.91 617	9	32
29	9.75 294	18	9.83 686	27	0.16 314	9.91 608	9	31
30	9.75 313	18	9.83 713	27	0.16 287	9.91 599	8	30
31	9.75 331	19	9.83 740	28	0.16 260	9.91 591	9	29
32	9.75 350	18	9.83 768	27	0.16 232	9.91 582	9	28
33	9.75 368	18	9.83 795	27	0.16 205	9.91 573	8	27
34	9.75 386	19	9.83 822	27	0.16 178	9.91 565	9	26
35	9.75 405	18	9.83 849	27	0.16 151	9.91 556	9	25
36	9.75 423	18	9.83 876	27	0.16 124	9.91 547	9	24
37	9.75 441	18	9.83 903	27	0.16 097	9.91 538	8	23
38	9.75 459	19	9.83 930	27	0.16 070	9.91 530	9	22
39	9.75 478	18	9.83 957	27	0.16 043	9.91 521	9	21
40	9.75 496	18	9.83 984	27	0.16 016	9.91 512	8	20
41	9.75 514	19	9.84 011	27	0.15 989	9.91 504	9	19
42	9.75 533	18	9.84 038	27	0.15 962	9.91 495	9	18
43	9.75 551	18	9.84 065	27	0.15 935	9.91 486	9	17
44	9.75 569	18	9.84 092	27	0.15 908	9.91 477	8	16
45	9.75 587	18	9.84 119	27	0.15 881	9.91 469	9	15
46	9.75 605	19	9.84 146	27	0.15 854	9.91 460	9	14
47	9.75 624	18	9.84 173	27	0.15 827	9.91 451	9	13
48	9.75 642	18	9.84 200	27	0.15 800	9.91 442	9	12
49	9.75 660	18	9.84 227	27	0.15 773	9.91 433	8	11
50	9.75 678	18	9.84 254	26	0.15 746	9.91 425	9	10
51	9.75 696	18	9.84 280	27	0.15 720	9.91 416	9	9
52	9.75 714	19	9.84 307	27	0.15 693	9.91 407	9	8
53	9.75 733	18	9.84 334	27	0.15 666	9.91 398	9	7
54	9.75 751	18	9.84 361	27	0.15 639	9.91 389	8	6
55	9.75 769	18	9.84 388	27	0.15 612	9.91 381	9	5
56	9.75 787	18	9.84 415	27	0.15 585	9.91 372	9	4
57	9.75 805	18	9.84 442	27	0.15 558	9.91 363	9	3
58	9.75 823	18	9.84 469	27	0.15 531	9.91 354	9	2
59	9.75 841	18	9.84 496	27	0.15 504	9.91 345	9	1
60	9.75 859		9.84 523		0.15 477	9.91 336		0
	L. Cos.	d.	L. Cotg.	c.d.	L. Tang.	L. Sin.	d.	′

P. P.

	28	27	26
1	0,5	0,4	0,4
2	0,9	0,9	0,9
3	1,4	1,4	1,3
4	1,9	1,8	1,7
5	2,3	2,2	2,2
6	2,8	2,7	2,6
7	3,3	3,2	3,0
8	3,7	3,6	3,5
9	4,2	4,0	3,9
10	4,7	4,5	4,3
20	9,3	9,0	8,7
30	14,0	13,5	13,0
40	18,7	18,0	17,3
50	23,3	22,5	21,7

	19	18
1	0,3	0,3
2	0,6	0,6
3	1,0	0,9
4	1,3	1,2
5	1,6	1,5
6	1,9	1,8
7	2,2	2,1
8	2,5	2,4
9	2,8	2,7
10	3,2	3,0
20	6,3	6,0
30	9,5	9,0
40	12,7	12,0
50	15,8	15,0

	9	8
1	0,2	0,1
2	0,3	0,3
3	0,4	0,4
4	0,6	0,5
5	0,8	0,7
6	0,9	0,9
7	1,0	0,9
8	1,2	1,1
9	1,4	1,2
10	1,5	1,3
20	3,0	2,7
30	4,5	4,0
40	6,0	5,3
50	7,5	6,7

55°

′	L. Sin.	d.	L. Tang.	c.d.	L. Cotg.	L. Cos.	d.		P. P.		
0	9.75 859	18	9.84 523	27	0.15 477	9.91 836	8	60			
1	9.75 877	18	9.84 550	26	0.15 450	9.91 828	9	59			
2	9.75 895	18	9.84 576	27	0.15 424	9.91 819	9	58			
3	9.75 913	18	9.84 603	27	0.15 397	9.91 810	9	57			
4	9.75 931	18	9.84 630	27	0.15 370	9.91 801	9	56			
5	9.75 949	18	9.84 657	27	0.15 343	9.91 292	9	55			
6	9.75 967	18	9.84 684	27	0.15 316	9.91 288	9	54			
7	9.75 985	18	9.84 711	27	0.15 289	9.91 274	8	53			
8	9.76 003	18	9.84 738	26	0.15 262	9.91 266	9	52			
9	9.76 021	18	9.84 764	27	0.15 236	9.91 257	9	51			
10	9.76 039	18	9.84 791	27	0.15 209	9.91 248	9	50	27	26	18
11	9.76 057	18	9.84 818	27	0.15 182	9.91 239	9	49			
12	9.76 075	18	9.84 845	27	0.15 155	9.91 230	9	48	1 0,4	0,4	0,3
13	9.76 093	18	9.84 872	27	0.15 128	9.91 221	9	47	2 0,9	0,9	0,6
14	9.76 111	18	9.84 899	26	0.15 101	9.91 212	9	46	3 1,4	1,3	0,9
									4 1,8	1,7	1,2
15	9.76 129	17	9.84 925	27	0.15 075	9.91 203	9	45			
16	9.76 146	18	9.84 952	27	0.15 048	9.91 194	9	44	5 2,2	2,2	1,5
17	9.76 164	18	9.84 979	27	0.15 021	9.91 185	9	43	6 2,7	2,6	1,8
18	9.76 182	18	9.85 006	27	0.14 994	9.91 176	9	42	7 3,2	3,0	2,1
19	9.76 200	18	9.85 033	26	0.14 967	9.91 167	9	41	8 3,6	3,5	2,4
									9 4,0	3,9	2,7
20	9.76 218	18	9.85 059	27	0.14 941	9.91 158	9	40			
21	9.76 236	17	9.85 086	27	0.14 914	9.91 149	8	39			
22	9.76 253	18	9.85 113	27	0.14 887	9.91 141	9	38	10 4,5	4,3	3,0
23	9.76 271	18	9.85 140	26	0.14 860	9.91 132	9	37	20 9,0	8,7	6,0
24	9.76 289	18	9.85 166	27	0.14 834	9.91 123	9	36	30 13,5	13,0	9,0
									40 18,0	17,3	12,0
25	9.76 307	17	9.85 193	27	0.14 807	9.91 114	9	35	50 22,5	21,7	15,0
26	9.76 324	18	9.85 220	27	0.14 780	9.91 105	9	34			
27	9.76 342	18	9.85 247	26	0.14 753	9.91 096	9	33			
28	9.76 360	18	9.85 273	27	0.14 727	9.91 087	9	32			
29	9.76 378	17	9.85 300	27	0.14 700	9.91 078	9	31			
30	9.76 395	18	9.85 327	27	0.14 673	9.91 069	9	30			
31	9.76 413	18	9.85 354	26	0.14 646	9.91 060	9	29			
32	9.76 431	17	9.85 380	27	0.14 620	9.91 051	9	28			
33	9.76 448	18	9.85 407	27	0.14 593	9.91 042	9	27			
34	9.76 466	18	9.85 434	26	0.14 566	9.91 033	10	26			
35	9.76 484	17	9.85 460	27	0.14 540	9.91 023	9	25			
36	9.76 501	18	9.85 487	27	0.14 513	9.91 014	9	24	17	10	9 8
37	9.76 519	18	9.85 514	26	0.14 486	9.91 005	9	23			
38	9.76 537	17	9.85 540	27	0.14 460	9.90 996	9	22	1 0,3	0,2	0,2 0,1
39	9.76 554	18	9.85 567	27	0.14 433	9.90 987	9	21	2 0,6	0,3	0,3 0,3
									3 0,8	0,5	0,4 0,4
40	9.76 572	18	9.85 594	26	0.14 406	9.90 978	9	20	4 1,1	0,7	0,6 0,5
41	9.76 590	17	9.85 620	27	0.14 380	9.90 969	9	19			
42	9.76 607	18	9.85 647	27	0.14 353	9.90 960	9	18	5 1,4	0,8	0,8 0,7
43	9.76 625	17	9.85 674	26	0.14 326	9.90 951	9	17	6 1,7	1,0	0,9 0,8
44	9.76 642	18	9.85 700	27	0.14 300	9.90 942	9	16	7 2,0	1,2	1,0 0,9
45	9.76 660	17	9.85 727	27	0.14 273	9.90 933	9	15	8 2,3	1,3	1,2 1,1
46	9.76 677	18	9.85 754	26	0.14 246	9.90 924	9	14	9 2,6	1,5	1,4 1,2
47	9.76 695	17	9.85 780	27	0.14 220	9.90 915	9	13			
48	9.76 712	18	9.85 807	27	0.14 193	9.90 906	10	12	10 2,8	1,7	1,5 1,3
49	9.76 730	17	9.85 834	26	0.14 166	9.90 896	9	11	20 5,7	3,3	3,0 2,7
									30 8,5	5,0	4,5 4,0
50	9.76 747	18	9.85 860	27	0.14 140	9.90 887	9	10	40 11,3	6,7	6,0 5,3
51	9.76 765	17	9.85 887	26	0.14 113	9.90 878	9	9	50 14,2	8,3	7,5 6,7
52	9.76 782	18	9.85 913	27	0.14 087	9.90 869	9	8			
53	9.76 800	17	9.85 940	27	0.14 060	9.90 860	9	7			
54	9.76 817	18	9.85 967	26	0.14 033	9.90 851	9	6			
55	9.76 835	17	9.85 993	27	0.14 007	9.90 842	10	5			
56	9.76 852	18	9.86 020	26	0.13 980	9.90 832	9	4			
57	9.76 870	17	9.86 046	27	0.13 954	9.90 823	9	3			
58	9.76 887	17	9.86 073	27	0.13 927	9.90 814	9	2			
59	9.76 904	18	9.86 100	26	0.13 900	9.90 805	9	1			
60	9.76 922		9.86 126		0.13 874	9.90 796		0			
	L. Cos.	d.	L. Cotg.	c.d.	L. Tang.	L. Sin.	d.	′		P. P.	

'	L. Sin.	d.	L. Tang.	c.d.	L. Cotg.	L. Cos.	d.		P. P.
0	9.76 922	17	9.86 126	27	0.13 874	9.90 796	9	60	
1	9.76 939	18	9.86 158	26	0.13 847	9.90 787	10	59	
2	9.76 957	17	9.86 179	27	0.13 821	9.90 777	10	58	
3	9.76 974	17	9.86 206	26	0.13 794	9.90 768	9	57	27 26
4	9.76 991	18	9.86 232	27	0.13 768	9.90 759	9	56	
5	9.77 009	17	9.86 259	26	0.13 741	9.90 750		55	1 0,4 0,4
6	9.77 026	17	9.86 285	27	0.13 715	9.90 741	10	54	2 0,9 0,9
7	9.77 043	18	9.86 312	26	0.13 688	9.90 731	9	53	3 1,4 1,3
8	9.77 061	17	9.86 338	27	0.13 662	9.90 722	9	52	4 1,8 1,7
9	9.77 078	17	9.86 365	27	0.13 635	9.90 713		51	
10	9.77 095	17	9.86 392	26	0.13 608	9.90 704	10	50	5 2,2 2,2
11	9.77 112	18	9.86 418	27	0.13 582	9.90 694	9	49	6 2,7 2,6
12	9.77 130	17	9.86 445	26	0.13 555	9.90 685	9	48	7 3,2 3,0
13	9.77 147	17	9.86 471	27	0.13 529	9.90 676	10	47	8 3,6 3,5
14	9.77 164	17	9.86 498	26	0.13 502	9.90 667	10	46	9 4,0 3,9
15	9.77 181	18	9.86 524	27	0.13 476	9.90 657	9	45	10 4,5 4,3
16	9.77 199	17	9.86 551	26	0.13 449	9.90 648	9	44	20 9,0 8,7
17	9.77 216	17	9.86 577	26	0.13 423	9.90 639	9	43	30 13,5 13,0
18	9.77 233	17	9.86 603	27	0.13 397	9.90 630	10	42	40 18,0 17,3
19	9.77 250	18	9.86 630	26	0.13 370	9.90 620	9	41	50 22,5 21,7
20	9.77 268	17	9.86 656	27	0.13 344	9.90 611	9	40	
21	9.77 285	17	9.86 683	26	0.13 317	9.90 602	10	39	
22	9.77 302	17	9.86 709	27	0.13 291	9.90 592	10	38	
23	9.77 319	17	9.86 736	26	0.13 264	9.90 583	9	37	18 17 16
24	9.77 336	17	9.86 762	27	0.13 238	9.90 574	9	36	
25	9.77 353	17	9.86 789	26	0.13 211	9.90 565	10	35	1 0,3 0,3 0,3
26	9.77 370	17	9.86 815	27	0.13 185	9.90 555	9	34	2 0,6 0,6 0,5
27	9.77 387	18	9.86 842	26	0.13 158	9.90 546	9	33	3 0,9 0,8 0,8
28	9.77 405	17	9.86 868	26	0.13 132	9.90 537	10	32	4 1,2 1,1 1,1
29	9.77 422	17	9.86 894	27	0.13 106	9.90 527	9	31	
30	9.77 439	17	9.86 921	26	0.13 079	9.90 518	9	30	5 1,5 1,4 1,3
31	9.77 456	17	9.86 947	27	0.13 053	9.90 509	10	29	6 1,8 1,7 1,6
32	9.77 473	17	9.86 974	26	0.13 026	9.90 499	9	28	7 2,1 2,0 1,9
33	9.77 490	17	9.87 000	27	0.13 000	9.90 490	10	27	8 2,4 2,3 2,1
34	9.77 507	17	9.87 027	26	0.12 973	9.90 480	9	26	9 2,7 2,6 2,4
35	9.77 524	17	9.87 053	26	0.12 947	9.90 471	9	25	10 3,0 2,8 2,7
36	9.77 541	17	9.87 079	27	0.12 921	9.90 462	10	24	20 6,0 5,7 5,3
37	9.77 558	17	9.87 106	26	0.12 894	9.90 452	9	23	30 9,0 8,5 8,0
38	9.77 575	17	9.87 132	26	0.12 868	9.90 443	9	22	40 12,0 11,3 10,7
39	9.77 592	17	9.87 158	27	0.12 842	9.90 434	10	21	50 15,0 14,2 13,3
40	9.77 609	17	9.87 185	26	0.12 815	9.90 424	9	20	
41	9.77 626	17	9.87 211	27	0.12 789	9.90 415	10	19	
42	9.77 643	17	9.87 238	26	0.12 762	9.90 405	10	18	10 . 9
43	9.77 660	17	9.87 264	26	0.12 736	9.90 396	10	17	
44	9.77 677	17	9.87 290	27	0.12 710	9.90 386	9	16	1 0,2 0,2
45	9.77 694	17	9.87 317	26	0.12 683	9.90 377	9	15	2 0,3 0,3
46	9.77 711	17	9.87 343	26	0.12 657	9.90 368	10	14	3 0,5 0,4
47	9.77 728	16	9.87 369	27	0.12 631	9.90 358	9	13	4 0,7 0,6
48	9.77 744	17	9.87 396	26	0.12 604	9.90 349	10	12	
49	9.77 761	17	9.87 422	26	0.12 578	9.90 339	10	11	5 0,8 0,8
50	9.77 778	17	9.87 448	27	0.12 552	9.90 330	10	10	6 1,0 0,9
51	9.77 795	17	9.87 475	26	0.12 525	9.90 320	9	9	7 1,2 1,0
52	9.77 812	17	9.87 501	26	0.12 499	9.90 311	10	8	8 1,3 1,2
53	9.77 829	17	9.87 527	27	0.12 473	9.90 301	10	7	9 1,5 1,4
54	9.77 846	16	9.87 554	26	0.12 446	9.90 292	10	6	10 1,7 1,5
55	9.77 862	17	9.87 580	26	0.12 420	9.90 282	9	5	20 3,3 3,0
56	9.77 879	17	9.87 606	27	0.12 394	9.90 273	10	4	30 5,0 4,5
57	9.77 896	17	9.87 633	26	0.12 367	9.90 263	9	3	40 6,7 6,0
58	9.77 913	17	9.87 659	26	0.12 341	9.90 254	10	2	50 8,3 7,5
59	9.77 930	16	9.87 685	26	0.12 315	9.90 244	10	1	
60	9.77 946		9.87 711		0.12 289	9.90 235		0	
	L. Cos.	d.	L. Cotg.	c.d.	L. Tang.	L. Sin.	d.	'	P. P.

′	L. Sin.	d.	L. Tang.	c.d.	L. Cotg.	L. Cos.	d.		P. P.	
0	9.77 946		9.87 711		0.12 289	9.90 285		60		
1	9.77 963	17	9.87 788	27	0.12 262	9.90 225	10	59		
2	9.77 980	17	9.87 764	26	0.12 236	9.90 216	9	58		
3	9.77 997	17	9.87 790	26	0.12 210	9.90 206	10	57		
4	9.78 013	16	9.87 817	27	0.12 188	9.90 197	9	56	27	26
		17		26			10			
5	9.78 030		9.87 843		0.12 157	9.90 187		55	1 0,4	0,4
6	9.78 047	17	9.87 869	26	0.12 131	9.90 178	9	54	2 0,9	0,9
7	9.78 063	16	9.87 895	26	0.12 105	9.90 168	10	53	3 1,4	1,3
8	9.78 080	17	9.87 922	27	0.12 078	9.90 159	9	52	4 1,8	1,7
9	9.78 097	17	9.87 948	26	0.12 052	9.90 149	10	51		
		16					10		5 2,2	2,2
10	9.78 118		9.87 974		0.12 026	9.90 139		50	6 2,7	2,6
11	9.78 130	17	9.88 000	26	0.12 000	9.90 130	10	49	7 3,2	3,0
12	9.78 147	17	9.88 027	27	0.11 973	9.90 120	10	48	8 3,6	3,5
13	9.78 163	16	9.88 053	26	0.11 947	9.90 111	9	47	9 4,0	3,9
14	9.78 180	17	9.88 079	26	0.11 921	9.90 101	10	46		
		17		26			10		10 4,5	4,3
15	9.78 197		9.88 105		0.11 895	9.90 091		45	20 9,0	8,7
16	9.78 213	16	9.88 131	26	0.11 869	9.90 082	9	44	30 13,5	13,0
17	9.78 230	17	9.88 158	27	0.11 842	9.90 072	10	43	40 18,0	17,3
18	9.78 246	16	9.88 184	26	0.11 816	9.90 063	9	42	50 22,5	21,7
19	9.78 263	17	9.88 210	26	0.11 790	9.90 053	10	41		
		17		26			10			
20	9.78 280		9.88 236		0.11 764	9.90 043		40		
21	9.78 296	16	9.88 262	26	0.11 738	9.90 034	9	39		
22	9.78 313	17	9.88 289	27	0.11 711	9.90 024	10	38		
23	9.78 329	16	9.88 315	26	0.11 685	9.90 014	10	37	17	16
24	9.78 346	17	9.88 341	26	0.11 659	9.90 005	9	36		
		16		26			10		1 0,3	0,3
25	9.78 362		9.88 367		0.11 633	9.89 995		35	2 0,6	0,5
26	9.78 379	17	9.88 393	26	0.11 607	9.89 985	10	34	3 0,8	0,8
27	9.78 395	16	9.88 420	27	0.11 580	9.89 976	9	33	4 1,1	1,1
28	9.78 412	17	9.88 446	26	0.11 554	9.89 966	10	32		
29	9.78 428	16	9.88 472	26	0.11 528	9.89 956	10	31	5 1,4	1,3
		17		26			9		6 1,7	1,6
30	9.78 445		9.88 498		0.11 502	9.89 947		30	7 2,0	1,9
31	9.78 461	16	9.88 524	26	0.11 476	9.89 937	10	29	8 2,3	2,1
32	9.78 478	17	9.88 550	26	0.11 450	9.89 927	10	28	9 2,6	2,4
33	9.78 494	16	9.88 577	27	0.11 423	9.89 918	9	27		
34	9.78 510	16	9.88 603	26	0.11 397	9.89 908	10	26	10 2,8	2,7
		17					10		20 5,7	5,3
35	9.78 527		9.88 629		0.11 371	9.89 898		25	30 8,5	8,0
36	9.78 543	16	9.88 655	26	0.11 345	9.89 888	10	24	40 11,3	10,7
37	9.78 560	17	9.88 681	26	0.11 319	9.89 879	9	23	50 14,2	13,3
38	9.78 576	16	9.88 707	26	0.11 293	9.89 869	10	22		
39	9.78 592	16	9.88 733	26	0.11 267	9.89 859	10	21		
		17		26			10			
40	9.78 609		9.88 759		0.11 241	9.89 849		20		
41	9.78 625	16	9.88 786	27	0.11 214	9.89 840	9	19		
42	9.78 642	17	9.88 812	26	0.11 188	9.89 830	10	18	10	9
43	9.78 658	16	9.88 838	26	0.11 162	9.89 820	10	17		
44	9.78 674	16	9.88 864	26	0.11 136	9.89 810	10	16	1 0,2	0,2
		17					9		2 0,3	0,3
45	9.78 691		9.88 890		0.11 110	9.89 801		15	3 0,5	0,4
46	9.78 707	16	9.88 916	26	0.11 084	9.89 791	10	14	4 0,7	0,6
47	9.78 723	16	9.88 942	26	0.11 058	9.89 781	10	13		
48	9.78 739	16	9.88 968	26	0.11 032	9.89 771	10	12	5 0,8	0,8
49	9.78 756	17	9.88 994	26	0.11 006	9.89 761	10	11	6 1,0	0,9
		16		26			9		7 1,2	1,0
50	9.78 772		9.89 020		0.10 980	9.89 752		10	8 1,3	1,2
51	9.78 788	16	9.89 046	26	0.10 954	9.89 742	10	9	9 1,5	1,4
52	9.78 805	17	9.89 073	27	0.10 927	9.89 732	10	8		
53	9.78 821	16	9.89 099	26	0.10 901	9.89 722	10	7	10 1,7	1,5
54	9.78 837	16	9.89 125	26	0.10 875	9.89 712	10	6	20 3,3	3,0
		16							30 5,0	4,5
55	9.78 853		9.89 151		0.10 849	9.89 702		5	40 6,7	6,0
56	9.78 869	16	9.89 177	26	0.10 823	9.89 693	9	4	50 8,3	7,5
57	9.78 886	17	9.89 203	26	0.10 797	9.89 683	10	3		
58	9.78 902	16	9.89 229	26	0.10 771	9.89 673	10	2		
59	9.78 918	16	9.89 255	26	0.10 745	9.89 663	10	1		
60	9.78 934		9.89 281		0.10 719	9.89 653		0		
	L. Cos.	d.	L. Cotg.	c.d.	L. Tang.	L. Sin.	d.	′	P. P.	

38°

'	L. Sin.	d.	L. Tang.	c.d.	L. Cotg.	L. Cos.	d.		P. P.
0	9.78 934	16	9.89 281	26	0.10 719	9.89 658	10	60	
1	9.78 950	17	9.89 307	26	0.10 693	9.89 648	10	59	
2	9.78 967	16	9.89 333	26	0.10 667	9.89 638	9	58	
3	9.78 983	16	9.89 359	26	0.10 641	9.89 624	10	57	
4	9.78 999	16	9.89 385	26	0.10 615	9.89 614	10	56	
									26 **25**
5	9.79 015	16	9.89 411	26	0.10 589	9.89 604	10	55	
6	9.79 031	16	9.89 437	26	0.10 563	9.89 594	10	54	1 | 0,4 0,4
7	9.79 047	16	9.89 463	26	0.10 537	9.89 584	10	53	2 | 0,9 0,8
8	9.79 063	16	9.89 489	26	0.10 511	9.89 574	10	52	3 | 1,3 1,2
9	9.79 079	16	9.89 515	26	0.10 485	9.89 564	10	51	4 | 1,7 1,7
10	9.79 095	16	9.89 541	26	0.10 459	9.89 554	10	50	5 | 2,2 2,1
11	9.79 111	17	9.89 567	26	0.10 433	9.89 544	10	49	6 | 2,6 2,5
12	9.79 128	16	9.89 593	26	0.10 407	9.89 534	10	48	7 | 3,0 2,9
13	9.79 144	16	9.89 619	26	0.10 381	9.89 524	10	47	8 | 3,5 3,3
14	9.79 160	16	9.89 645	26	0.10 355	9.89 514	10	46	9 | 3,9 3,8
15	9.79 176	16	9.89 671	26	0.10 329	9.89 504	9	45	10 | 4,3 4,2
16	9.79 192	16	9.89 697	26	0.10 303	9.89 495	10	44	20 | 8,7 8,3
17	9.79 208	16	9.89 723	26	0.10 277	9.89 485	10	43	30 | 13,0 12,5
18	9.79 224	16	9.89 749	26	0.10 251	9.89 475	10	42	40 | 17,3 16,7
19	9.79 240	16	9.89 775	26	0.10 225	9.89 465	10	41	50 | 21,7 20,8
20	9.79 256	16	9.89 801	26	0.10 199	9.89 455	10	40	
21	9.79 272	16	9.89 827	26	0.10 173	9.89 445	10	39	
22	9.79 288	16	9.89 853	26	0.10 147	9.89 435	10	38	
23	9.79 304	16	9.89 879	26	0.10 121	9.89 425	10	37	**17** **16** **15**
24	9.79 319	15	9.89 905	26	0.10 095	9.89 415	10	36	
25	9.79 335	16	9.89 931	26	0.10 069	9.89 405	10	35	1 | 0,3 0,3 0,2
26	9.79 351	16	9.89 957	26	0.10 043	9.89 395	10	34	2 | 0,6 0,5 0,5
27	9.79 367	16	9.89 983	26	0.10 017	9.89 385	10	33	3 | 0,8 0,8 0,8
28	9.79 383	16	9.90 009	26	0.09 991	9.89 375	11	32	4 | 1,1 1,1 1,0
29	9.79 399	16	9.90 035	26	0.09 965	9.89 364	10	31	5 | 1,4 1,3 1,2
30	9.79 415	16	9.90 061	25	0.09 939	9.89 354	10	30	6 | 1,7 1,6 1,5
31	9.79 431	16	9.90 086	26	0.09 914	9.89 344	10	29	7 | 2,0 1,9 1,8
32	9.79 447	16	9.90 112	26	0.09 888	9.89 334	10	28	8 | 2,3 2,1 2,0
33	9.79 463	15	9.90 138	26	0.09 862	9.89 324	10	27	9 | 2,6 2,4 2,2
34	9.79 478	16	9.90 164	26	0.09 836	9.89 314	10	26	10 | 2,8 2,7 2,5
35	9.79 494	16	9.90 190	26	0.09 810	9.89 304	10	25	20 | 5,7 5,3 5,0
36	9.79 510	16	9.90 216	26	0.09 784	9.89 294	10	24	30 | 8,5 8,0 7,5
37	9.79 526	16	9.90 242	26	0.09 758	9.89 284	10	23	40 | 11,3 10,7 10,0
38	9.79 542	16	9.90 268	26	0.09 732	9.89 274	10	22	50 | 14,2 13,3 12,5
39	9.79 558	15	9.90 294	26	0.09 706	9.89 264	10	21	
40	9.79 573	16	9.90 320	26	0.09 680	9.89 254	10	20	
41	9.79 589	16	9.90 346	25	0.09 654	9.89 244	11	19	
42	9.79 605	16	9.90 371	26	0.09 629	9.89 233	10	18	**11** **10** **9**
43	9.79 621	15	9.90 397	26	0.09 603	9.89 223	10	17	
44	9.79 636	16	9.90 423	26	0.09 577	9.89 213	10	16	
45	9.79 652	16	9.90 449	26	0.09 551	9.89 203	10	15	1 | 0,2 0,2 0,2
46	9.79 668	16	9.90 475	26	0.09 525	9.89 193	10	14	2 | 0,4 0,3 0,3
47	9.79 684	15	9.90 501	26	0.09 499	9.89 183	10	13	3 | 0,6 0,5 0,4
48	9.79 699	16	9.90 527	26	0.09 473	9.89 173	11	12	4 | 0,7 0,7 0,6
49	9.79 715	16	9.90 553	25	0.09 447	9.89 162	11	11	5 | 0,9 0,8 0,8
50	9.79 731	15	9.90 578	26	0.09 422	9.89 152	10	10	6 | 1,1 1,0 0,9
51	9.79 746	16	9.90 604	26	0.09 396	9.89 142	10	9	7 | 1,3 1,2 1,0
52	9.79 762	16	9.90 630	26	0.09 370	9.89 132	10	8	8 | 1,5 1,3 1,2
53	9.79 778	15	9.90 656	26	0.09 344	9.89 122	10	7	9 | 1,6 1,5 1,4
54	9.79 793	16	9.90 682	26	0.09 318	9.89 112	11	6	10 | 1,8 1,7 1,5
55	9.79 809	16	9.90 708	26	0.09 292	9.89 101	10	5	20 | 3,7 3,3 3,0
56	9.79 825	15	9.90 734	25	0.09 266	9.89 091	10	4	30 | 5,5 5,0 4,5
57	9.79 840	16	9.90 759	26	0.09 241	9.89 081	10	3	40 | 7,3 6,7 6,0
58	9.79 856	16	9.90 785	26	0.09 215	9.89 071	11	2	50 | 9,2 8,3 7,5
59	9.79 872	15	9.90 811	26	0.09 189	9.89 060	10	1	
60	9.79 887		9.90 837		0.09 163	9.89 050		0	
	L. Cos.	d.	L. Cotg.	c.d.	L. Tang.	L. Sin.	d.	'	P. P.

51°

39°

'	L. Sin.	d.	L. Tang.	c.d.	L. Cotg.	L. Cos.	d.		P. P.	
0	9.79 887	16	9.90 887	26	0.09 163	9.89 050	10	60		
1	9.79 903	15	9.90 863	26	0.09 137	9.89 040	10	59		
2	9.79 918	16	9.90 889	25	0.09 111	9.89 030	10	58		
3	9.79 934	16	9.90 914	26	0.09 086	9.89 020	11	57		
4	9.79 950	15	9.90 940	26	0.09 060	9.89 009	10	56	**26**	**25**
5	9.79 965	16	9.90 966	26	0.09 034	9.88 999	10	55	1	0,4	0,4
6	9.79 981	15	9.90 992	26	0.09 008	9.88 989	11	54	2	0,9	0,8
7	9.79 996	16	9.91 018	25	0.08 982	9.88 978	10	53	3	1,3	1,2
8	9.80 012	15	9.91 043	26	0.08 957	9.88 968	10	52	4	1,7	1,7
9	9.80 027	16	9.91 069	26	0.08 931	9.88 958	10	51		
10	9.80 043	15	9.91 095	26	0.08 905	9.88 948	11	50	5	2,2	2,1
11	9.80 058	16	9.91 121	26	0.08 879	9.88 937	10	49	6	2,6	2,5
12	9.80 074	15	9.91 147	25	0.08 853	9.88 927	10	48	7	3,0	2,9
13	9.80 089	16	9.91 172	26	0.08 828	9.88 917	11	47	8	3,5	3,3
14	9.80 105	15	9.91 198	26	0.08 802	9.88 906	10	46	9	3,9	3,8
15	9.80 120	16	9.91 224	26	0.08 776	9.88 896	10	45	10	4,3	4,2
16	9.80 136	15	9.91 250	26	0.08 750	9.88 886	11	44	20	8,7	8,3
17	9.80 151	15	9.91 276	25	0.08 724	9.88 875	10	43	30	13,0	12,5
18	9.80 166	16	9.91 301	26	0.08 699	9.88 865	10	42	40	17,3	16,7
19	9.80 182	15	9.91 327	26	0.08 673	9.88 855	11	41	50	21,7	20,8
20	9.80 197	16	9.91 353	26	0.08 647	9.88 844	10	40		
21	9.80 213	15	9.91 379	25	0.08 621	9.88 834	10	39		
22	9.80 228	16	9.91 404	26	0.08 596	9.88 824	11	38		
23	9.80 244	15	9.91 430	26	0.08 570	9.88 813	10	37	**16**	**15**
24	9.80 259	15	9.91 456	26	0.08 544	9.88 803	10	36	1	0,3	0,2
25	9.80 274	16	9.91 482	25	0.08 518	9.88 793	11	35	2	0,5	0,5
26	9.80 290	15	9.91 507	26	0.08 493	9.88 782	10	34	3	0,8	0,8
27	9.80 305	15	9.91 533	26	0.08 467	9.88 772	11	33	4	1,1	1,0
28	9.80 320	16	9.91 559	26	0.08 441	9.88 761	10	32		
29	9.80 336	15	9.91 585	25	0.08 415	9.88 751	10	31	5	1,3	1,2
									6	1,6	1,5
30	9.80 351	15	9.91 610	26	0.08 390	9.88 741	11	30	7	1,9	1,8
31	9.80 366	16	9.91 636	26	0.08 364	9.88 730	10	29	8	2,1	2,0
32	9.80 382	15	9.91 662	26	0.08 338	9.88 720	11	28	9	2,4	2,2
33	9.80 397	15	9.91 688	25	0.08 312	9.88 709	10	27		
34	9.80 412	16	9.91 713	26	0.08 287	9.88 699	11	26	10	2,7	2,5
35	9.80 428	15	9.91 739	26	0.08 261	9.88 688	10	25	20	5,3	5,0
36	9.80 443	15	9.91 765	26	0.08 235	9.88 678	10	24	30	8,0	7,5
37	9.80 458	15	9.91 791	25	0.08 209	9.88 668	11	23	40	10,7	10,0
38	9.80 473	16	9.91 816	26	0.08 184	9.88 657	10	22	50	13,3	12,5
39	9.80 489	15	9.91 842	26	0.08 158	9.88 647	11	21		
40	9.80 504	15	9.91 868	25	0.08 132	9.88 636	10	20		
41	9.80 519	15	9.91 893	26	0.08 107	9.88 626	11	19		
42	9.80 534	16	9.91 919	26	0.08 081	9.88 615	10	18	**11**	**10**
43	9.80 550	15	9.91 945	26	0.08 055	9.88 605	11	17	1	0,2	0,2
44	9.80 565	15	9.91 971	25	0.08 029	9.88 594	10	16	2	0,4	0,3
45	9.80 580	15	9.91 996	26	0.08 004	9.88 584	11	15	3	0,6	0,5
46	9.80 595	15	9.92 022	26	0.07 978	9.88 573	10	14	4	0,7	0,7
47	9.80 610	15	9.92 048	25	0.07 952	9.88 563	11	13		
48	9.80 625	16	9.92 073	26	0.07 927	9.88 552	10	12	5	0,9	0,8
49	9.80 641	15	9.92 099	26	0.07 901	9.88 542	11	11	6	1,1	1,0
									7	1,3	1,2
50	9.80 656	15	9.92 125	25	0.07 875	9.88 531	10	10	8	1,5	1,3
51	9.80 671	15	9.92 150	26	0.07 850	9.88 521	11	9	9	1,6	1,5
52	9.80 686	15	9.92 176	26	0.07 824	9.88 510	11	8		
53	9.80 701	15	9.92 202	25	0.07 798	9.88 499	10	7	10	1,8	1,7
54	9.80 716	15	9.92 227	26	0.07 773	9.88 489	11	6	20	3,7	3,3
55	9.80 731	15	9.92 253	26	0.07 747	9.88 478	10	5	30	5,5	5,0
56	9.80 746	16	9.92 279	25	0.07 721	9.88 468	11	4	40	7,3	6,7
57	9.80 762	15	9.92 304	26	0.07 696	9.88 457	10	3	50	9,2	8,3
58	9.80 777	15	9.92 330	26	0.07 670	9.88 447	11	2		
59	9.80 792	15	9.92 356	25	0.07 644	9.88 436	11	1		
60	9.80 807		9.92 381		0.07 619	9.88 425		0		
	L. Cos.	d.	L. Cotg.	c.d.	L. Tang.	L. Sin.	d.	'	P. P.	

50°

9.8 1036

'	L. Sin.	d.	L. Tang.	c.d.	L. Cotg.	L. Cos.	d.		P. P.	
0	9.80 807	15	9.92 381	26	0.07 619	9.88 425	10	**60**		
1	9.80 822	15	9.92 407	26	0.07 598	9.88 415	11	59		
2	9.80 837	15	9.92 438	25	0.07 567	9.88 404	10	58		
8	9.80 852	15	9.92 458	26	0.07 542	9.88 894	11	57		
4	9.80 867	15	9.92 484	26	0.07 516	9.88 383	11	56	**26**	**25**
5	9.80 882	15	9.92 510	25	0.07 490	9.88 372	10	55	1	0,4	0,4
6	9.80 897	15	9.92 585	26	0.07 465	9.88 362	11	54	2	0,9	0,8
7	9.80 912	15	9.92 561	26	0.07 439	9.88 851	11	53	8	1,8	1,2
8	9.80 927	15	9.92 587	25	0.07 413	9.88 340	10	52	4	1,7	1,7
9	9.80 942	15	9.92 612	26	0.07 388	9.88 330	11	51		
10	9.80 957	15	9.92 638	25	0.07 362	9.88 319	11	**50**	5	2,3	2,1
11	9.80 972	15	9.92 663	26	0.07 337	9.88 308	10	49	6	2,6	2,5
12	9.80 987	15	9.92 689	26	0.07 311	9.88 298	11	48	7	8,0	2,9
18	9.81 002	15	9.92 715	25	0.07 285	9.88 287	11	47	8	8,5	8,8
14	9.81 017	15	9.92 740	26	0.07 260	9.88 276	10	46	9	8,9	8,8
15	9.81 032	15	9.92 766	26	0.07 234	9.88 266	11	**45**	10	4,3	4,2
16	9.81 047	14	9.92 792	25	0.07 208	9.88 255	11	44	20	8,7	8,8
17	9.81 061	15	9.92 817	26	0.07 183	9.88 244	10	43	30	13,0	12,5
18	9.81 076	15	9.92 843	25	0.07 157	9.88 234	11	42	40	17,8	16,7
19	9.81 091	15	9.92 868	26	0.07 132	9.88 223	11	41	50	21,7	20,8
20	9.81 106	15	9.92 894	26	0.07 106	9.88 212	11	**40**		
21	9.81 121	15	9.92 920	25	0.07 080	9.88 201	10	39		
22	9.81 136	15	9.92 945	26	0.07 055	9.88 191	11	38		
28	9.81 151	15	9.92 971	25	0.07 029	9.88 180	11	87	**15**	**14**
24	9.81 166	14	9.92 996	26	0.07 004	9.88 169	11	36	1	0,2	0,2
25	9.81 180	15	9.93 022	26	0.06 978	9.88 158	10	**35**	2	0,5	0,5
26	9.81 195	15	9.93 048	25	0.06 952	9.88 148	11	34	8	0,8	0,7
27	9.81 210	15	9.93 073	26	0.06 927	9.88 137	11	33	4	1,0	0,9
28	9.81 225	15	9.93 099	25	0.06 901	9.88 126	11	32		
29	9.81 240	14	9.93 124	26	0.06 876	9.88 115	10	31	5	1,2	1,2
									6	1,5	1,4
30	9.81 254	15	9.93 150	25	0.06 850	9.88 105	11	**30**	7	1,8	1,6
81	9.81 269	15	9.93 175	26	0.06 825	9.88 094	11	29	8	2,0	1,9
82	9.81 284	15	9.93 201	26	0.06 799	9.88 083	11	28	9	2,2	2,1
88	9.81 299	15	9.93 227	25	0.06 773	9.88 072	11	27		
84	9.81 314	14	9.93 252	26	0.06 748	9.88 061	10	26	10	2,5	2,3
85	9.81 328	15	9.93 278	25	0.06 722	9.88 051	11	**25**	20	5,0	4,7
86	9.81 343	15	9.93 303	26	0.06 697	9.88 040	11	24	30	7,5	7,0
87	9.81 358	14	9.93 329	25	0.06 671	9.88 029	11	28	40	10,0	9,3
88	9.81 372	15	9.93 354	26	0.06 646	9.88 018	11	22	50	12,5	11,7
89	9.81 387	15	9.93 380	26	0.06 620	9.88 007	11	21		
40	9.81 402	15	9.93 406	25	0.06 594	9.87 996	11	**20**		
41	9.81 417	14	9.93 431	26	0.06 569	9.87 985	10	19		
42	9.81 431	15	9.93 457	25	0.06 543	9.87 975	11	18	**11**	**10**
48	9.81 446	15	9.93 482	26	0.06 518	9.87 964	11	17	1	0,2	0,2
44	9.81 461	14	9.93 508	25	0.06 492	9.87 953	11	16	2	0,4	0,3
45	9.81 475	15	9.93 533	26	0.06 467	9.87 942	11	**15**	8	0,6	0,5
46	9.81 490	15	9.93 559	25	0.06 441	9.87 931	11	14	4	0,7	0,7
47	9.81 505	14	9.93 584	26	0.06 416	9.87 920	11	18		
48	9.81 519	15	9.93 610	26	0.06 390	9.87 909	11	12	5	0,9	0,8
49	9.81 534	15	9.93 636	25	0.06 364	9.87 898	11	11	6	1,1	1,0
									7	1,3	1,2
50	9.81 549	14	9.93 661	26	0.06 389	9.87 887	10	**10**	8	1,5	1,3
51	9.81 563	15	9.93 687	25	0.06 313	9.87 877	11	9	9	1,6	1,5
52	9.81 578	14	9.93 712	26	0.06 288	9.87 866	11	8		
58	9.81 592	15	9.93 738	25	0.06 262	9.87 855	11	7	10	1,8	1,7
54	9.81 607	15	9.93 763	26	0.06 237	9.87 844	11	6	20	8,7	8,8
55	9.81 622	14	9.93 789	25	0.06 211	9.87 833	11	**5**	30	5,5	5,0
56	9.81 636	15	9.93 814	26	0.06 186	9.87 822	11	4	40	7,8	6,7
57	9.81 651	14	9.93 840	25	0.06 160	9.87 811	11	8	50	9,2	8,8
58	9.81 665	15	9.93 865	26	0.06 135	9.87 800	11	2		
59	9.81 680	14	9.93 891	25	0.06 109	9.87 789	11	1		
60	9.81 694		9.93 916		0.06 084	9.87 778		**0**		
	L. Cos.	d.	L. Cotg.	c.d.	L. Tang.	L. Sin.	d.	'		P. P.	

41°

′	L. Sin.	d.	L. Tang.	c.d.	L. Cotg.	L. Cos.	d.	
0	9.81 694	15	9.93 916	26	0.06 084	9.87 778	11	60
1	9.81 709	14	9.93 942	25	0.06 058	9.87 767	11	59
2	9.81 723	15	9.93 967	26	0.06 033	9.87 756	11	58
3	9.81 738	14	9.93 993	25	0.06 007	9.87 745	11	57
4	9.81 752	15	9.94 018	26	0.05 982	9.87 734	11	56
5	9.81 767	14	9.94 044	25	0.05 956	9.87 723	11	55
6	9.81 781	15	9.94 069	26	0.05 931	9.87 712	11	54
7	9.81 796	14	9.94 095	25	0.05 905	9.87 701	11	53
8	9.81 810	15	9.94 120	26	0.05 880	9.87 690	11	52
9	9.81 825	14	9.94 146	25	0.05 854	9.87 679	11	51
10	9.81 839	15	9.94 171	26	0.05 829	9.87 668	11	50
11	9.81 854	14	9.94 197	25	0.05 803	9.87 657	11	49
12	9.81 868	14	9.94 222	26	0.05 778	9.87 646	11	48
13	9.81 882	15	9.94 248	25	0.05 752	9.87 635	11	47
14	9.81 897	14	9.94 273	26	0.05 727	9.87 624	11	46
15	9.81 911	15	9.94 299	25	0.05 701	9.87 613	12	45
16	9.81 926	14	9.94 324	26	0.05 676	9.87 601	11	44
17	9.81 940	15	9.94 350	25	0.05 650	9.87 590	11	43
18	9.81 955	14	9.94 375	26	0.05 625	9.87 579	11	42
19	9.81 969	14	9.94 401	25	0.05 599	9.87 568	11	41
20	9.81 983	15	9.94 426	26	0.05 574	9.87 557	11	40
21	9.81 998	14	9.94 452	25	0.05 548	9.87 546	11	39
22	9.82 012	14	9.94 477	26	0.05 523	9.87 535	11	38
23	9.82 026	15	9.94 503	25	0.05 497	9.87 524	11	37
24	9.82 041	14	9.94 528	26	0.05 472	9.87 513	12	36
25	9.82 055	14	9.94 554	25	0.05 446	9.87 501	11	35
26	9.82 069	15	9.94 579	25	0.05 421	9.87 490	11	34
27	9.82 084	14	9.94 604	26	0.05 396	9.87 479	11	33
28	9.82 098	14	9.94 630	25	0.05 370	9.87 468	11	32
29	9.82 112	14	9.94 655	26	0.05 345	9.87 457	11	31
30	9.82 126	15	9.94 681	25	0.05 319	9.87 446	12	30
31	9.82 141	14	9.94 706	26	0.05 294	9.87 434	11	29
32	9.82 155	14	9.94 732	25	0.05 268	9.87 423	11	28
33	9.82 169	15	9.94 757	26	0.05 243	9.87 412	11	27
34	9.82 184	14	9.94 783	25	0.05 217	9.87 401	11	26
35	9.82 198	14	9.94 808	26	0.05 192	9.87 390	12	25
36	9.82 212	14	9.94 834	25	0.05 166	9.87 378	11	24
37	9.82 226	14	9.94 859	25	0.05 141	9.87 367	11	23
38	9.82 240	15	9.94 884	26	0.05 116	9.87 356	11	22
39	9.82 255	14	9.94 910	25	0.05 090	9.87 345	11	21
40	9.82 269	14	9.94 935	26	0.05 065	9.87 334	12	20
41	9.82 283	14	9.94 961	25	0.05 039	9.87 322	11	19
42	9.82 297	14	9.94 986	26	0.05 014	9.87 311	11	18
43	9.82 311	15	9.95 012	25	0.04 988	9.87 300	12	17
44	9.82 326	14	9.95 037	25	0.04 963	9.87 288	11	16
45	9.82 340	14	9.95 062	26	0.04 938	9.87 277	11	15
46	9.82 354	14	9.95 088	25	0.04 912	9.87 266	11	14
47	9.82 368	14	9.95 113	26	0.04 887	9.87 255	12	13
48	9.82 382	14	9.95 139	25	0.04 861	9.87 243	11	12
49	9.82 396	14	9.95 164	26	0.04 836	9.87 232	11	11
50	9.82 410	14	9.95 190	25	0.04 810	9.87 221	12	10
51	9.82 424	15	9.95 215	25	0.04 785	9.87 209	11	9
52	9.82 439	14	9.95 240	26	0.04 760	9.87 198	11	8
53	9.82 453	14	9.95 266	25	0.04 734	9.87 187	12	7
54	9.82 467	14	9.95 291	26	0.04 709	9.87 175	11	6
55	9.82 481	14	9.95 317	25	0.04 683	9.87 164	11	5
56	9.82 495	14	9.95 342	26	0.04 658	9.87 153	12	4
57	9.82 509	14	9.95 368	25	0.04 632	9.87 141	11	3
58	9.82 523	14	9.95 393	25	0.04 607	9.87 130	11	2
59	9.82 537	14	9.95 418	26	0.04 582	9.87 119	12	1
60	9.82 551		9.95 444		0.04 556	9.87 107		0
	L. Cos.	d.	L. Cotg.	c.d.	L. Tang.	L. Sin.	d.	′

P. P.

	26	25
1	0,4	0,4
2	0,9	0,8
3	1,3	1,2
4	1,7	1,7
5	2,2	2,1
6	2,6	2,5
7	3,0	2,9
8	3,5	3,3
9	3,9	3,8
10	4,3	4,2
20	8,7	8,3
30	13,0	12,5
40	17,3	16,7
50	21,7	20,8

	15	14
1	0,2	0,2
2	0,5	0,5
3	0,8	0,7
4	1,0	0,9
5	1,2	1,2
6	1,5	1,4
7	1,8	1,6
8	2,0	1,9
9	2,2	2,1
10	2,5	2,3
20	5,0	4,7
30	7,5	7,0
40	10,0	9,3
50	12,5	11,7

	12	11
1	0,2	0,2
2	0,4	0,4
3	0,6	0,6
4	0,8	0,7
5	1,0	0,9
6	1,2	1,1
7	1,4	1,3
8	1,6	1,5
9	1,8	1,6
10	2,0	1,8
20	4,0	3,7
30	6,0	5,5
40	8,0	7,3
50	10,0	9,2

48°

'	L. Sin.	d.	L. Tang.	c.d.	L. Cotg.	L. Cos.	d.	'
0	9.82 551	14	9.95 444	25	0.04 556	9.87 107	11	60
1	9.82 565	14	9.95 469	26	0.04 531	9.87 096	11	59
2	9.82 579	14	9.95 495	25	0.04 505	9.87 085	12	58
3	9.82 593	14	9.95 520	25	0.04 480	9.87 073	11	57
4	9.82 607	14	9.95 545	26	0.04 455	9.87 062	12	56
5	9.82 621	14	9.95 571	25	0.04 429	9.87 050	11	55
6	9.82 635	14	9.95 596	26	0.04 404	9.87 039	11	54
7	9.82 649	14	9.95 622	25	0.04 378	9.87 028	12	53
8	9.82 663	14	9.95 647	25	0.04 353	9.87 016	12	52
9	9.82 677	14	9.95 672	26	0.04 328	9.87 005	12	51
10	9.82 691	14	9.95 698	25	0.04 302	9.86 993	11	50
11	9.82 705	14	9.95 723	25	0.04 277	9.86 982	12	49
12	9.82 719	14	9.95 748	26	0.04 252	9.86 970	11	48
13	9.82 733	14	9.95 774	25	0.04 226	9.86 959	11	47
14	9.82 747	14	9.95 799	26	0.04 201	9.86 947	11	46
15	9.82 761	14	9.95 825	25	0.04 175	9.86 936	12	45
16	9.82 775	13	9.95 850	25	0.04 150	9.86 924	11	44
17	9.82 788	14	9.95 875	26	0.04 125	9.86 913	11	43
18	9.82 802	14	9.95 901	25	0.04 099	9.86 902	12	42
19	9.82 816	14	9.95 926	26	0.04 074	9.86 890	11	41
20	9.82 830	14	9.95 952	25	0.04 048	9.86 879	12	40
21	9.82 844	14	9.95 977	25	0.04 023	9.86 867	12	39
22	9.82 858	14	9.96 002	26	0.03 998	9.86 855	11	38
23	9.82 872	14	9.96 028	25	0.03 972	9.86 844	12	37
24	9.82 885	14	9.96 053	25	0.03 947	9.86 832	11	36
25	9.82 899	14	9.96 078	26	0.03 922	9.86 821	12	35
26	9.82 913	14	9.96 104	25	0.03 896	9.86 809	12	34
27	9.82 927	14	9.96 129	26	0.03 871	9.86 798	12	33
28	9.82 941	14	9.96 155	25	0.03 845	9.86 786	11	32
29	9.82 955	13	9.96 180	25	0.03 820	9.86 775	12	31
30	9.82 968	14	9.96 205	26	0.03 795	9.86 763		30
31	9.82 982	14	9.96 231	25	0.03 769	9.86 752	12	29
32	9.82 996	14	9.96 256	25	0.03 744	9.86 740	12	28
33	9.83 010	13	9.96 281	26	0.03 719	9.86 728	11	27
34	9.83 023	14	9.96 307	25	0.03 693	9.86 717	12	26
35	9.83 037	14	9.96 332	25	0.03 668	9.86 705	11	25
36	9.83 051	14	9.96 357	26	0.03 643	9.86 694	12	24
37	9.83 065	13	9.96 383	25	0.03 617	9.86 682	12	23
38	9.83 078	14	9.96 408	25	0.03 592	9.86 670	11	22
39	9.83 092	14	9.96 433	26	0.03 567	9.86 659	12	21
40	9.83 106	14	9.96 459	25	0.03 541	9.86 647	12	20
41	9.83 120	13	9.96 484	26	0.03 516	9.86 635	11	19
42	9.83 133	14	9.96 510	25	0.03 490	9.86 624	12	18
43	9.83 147	14	9.96 535	25	0.03 465	9.86 612	12	17
44	9.83 161	13	9.96 560	26	0.03 440	9.86 600	11	16
45	9.83 174	14	9.96 586	25	0.03 414	9.86 589	12	15
46	9.83 188	14	9.96 611	25	0.03 389	9.86 577	12	14
47	9.83 202	13	9.96 636	26	0.03 364	9.86 565	11	13
48	9.83 215	14	9.96 662	25	0.03 338	9.86 554	12	12
49	9.83 229	13	9.96 687	25	0.03 313	9.86 542	12	11
50	9.83 242	14	9.96 712	26	0.03 288	9.86 530		10
51	9.83 256	14	9.96 738	25	0.03 262	9.86 518	11	9
52	9.83 270	13	9.96 763	25	0.03 237	9.86 507	12	8
53	9.83 283	14	9.96 788	26	0.03 212	9.86 495	12	7
54	9.83 297	13	9.96 814	25	0.03 186	9.86 483	11	6
55	9.83 310	14	9.96 839	25	0.03 161	9.86 472	12	5
56	9.83 324	14	9.96 864	26	0.03 136	9.86 460	12	4
57	9.83 338	13	9.96 890	25	0.03 110	9.86 448	12	3
58	9.83 351	14	9.96 915	25	0.03 085	9.86 436	12	2
59	9.83 365	13	9.96 940	26	0.03 060	9.86 425	12	1
60	9.83 378		9.96 966		0.03 034	9.86 413		0
	L. Cos.	d.	L. Cotg.	c.d.	L. Tang.	L. Sin.	d.	'

P. P.

	26	25
1	0,4	0,4
2	0,9	0,8
3	1,3	1,2
4	1,7	1,7
5	2,2	2,1
6	2,6	2,5
7	3,0	2,9
8	3,5	3,3
9	3,9	3,8
10	4,3	4,2
20	8,7	8,3
30	13,0	12,5
40	17,3	16,7
50	21,7	20,8

	14	13
1	0,2	0,2
2	0,5	0,4
3	0,7	0,6
4	0,9	0,9
5	1,2	1,1
6	1,4	1,3
7	1,6	1,5
8	1,9	1,7
9	2,1	2,0
10	2,3	2,2
20	4,7	4,3
30	7,0	6,5
40	9,3	8,7
50	11,7	10,8

	12	11
1	0,2	0,2
2	0,4	0,4
3	0,6	0,6
4	0,8	0,7
5	1,0	0,9
6	1,2	1,1
7	1,4	1,3
8	1,6	1,5
9	1,8	1,6
10	2,0	1,8
20	4,0	3,7
30	6,0	5,5
40	8,0	7,3
50	10,0	9,2

′	L. Sin.	d.	L. Tang.	c.d.	L. Cotg.	L. Cos.	d.	′
0	9.83 878	14	9.96 966	25	0.03 034	9.86 418	12	60
1	9.83 392	13	9.96 991	25	0.03 009	9.86 401	12	59
2	9.83 405	14	9.97 016	26	0.02 984	9.86 389	12	58
3	9.83 419	13	9.97 042	25	0.02 958	9.86 377	11	57
4	9.83 432	14	9.97 067	25	0.02 933	9.86 366	12	56
5	9.83 446	13	9.97 092	26	0.02 908	9.86 354	12	55
6	9.83 459	14	9.97 118	25	0.02 882	9.86 342	12	54
7	9.83 473	13	9.97 143	25	0.02 857	9.86 330	12	53
8	9.83 486	14	9.97 168	25	0.02 832	9.86 318	12	52
9	9.83 500	13	9.97 193	26	0.02 807	9.86 306	11	51
10	9.83 513	14	9.97 219	25	0.02 781	9.86 295	12	50
11	9.83 527	13	9.97 244	25	0.02 756	9.86 283	12	49
12	9.83 540	14	9.97 269	26	0.02 731	9.86 271	12	48
13	9.83 554	13	9.97 295	25	0.02 705	9.86 259	12	47
14	9.83 567	14	9.97 320	25	0.02 680	9.86 247	12	46
15	9.83 581	13	9.97 345	26	0.02 655	9.86 235	12	45
16	9.83 594	14	9.97 371	25	0.02 629	9.86 223	12	44
17	9.83 608	13	9.97 396	25	0.02 604	9.86 211	11	43
18	9.83 621	13	9.97 421	26	0.02 579	9.86 200	12	42
19	9.83 634	14	9.97 447	25	0.02 553	9.86 188	12	41
20	9.83 648	13	9.97 472	25	0.02 528	9.86 176	12	40
21	9.83 661	13	9.97 497	26	0.02 503	9.86 164	12	39
22	9.83 674	14	9.97 523	25	0.02 477	9.86 152	12	38
23	9.83 688	13	9.97 548	25	0.02 452	9.86 140	12	37
24	9.83 701	14	9.97 573	25	0.02 427	9.86 128	12	36
25	9.83 715	13	9.97 598	26	0.02 402	9.86 116	12	35
26	9.83 728	13	9.97 624	25	0.02 376	9.86 104	12	34
27	9.83 741	14	9.97 649	25	0.02 351	9.86 092	12	33
28	9.83 755	13	9.97 674	26	0.02 326	9.86 080	12	32
29	9.83 768	13	9.97 700	25	0.02 300	9.86 068	12	31
30	9.83 781	14	9.97 725	25	0.02 275	9.86 056	12	30
31	9.83 795	13	9.97 750	26	0.02 250	9.86 044	12	29
32	9.83 808	13	9.97 776	25	0.02 224	9.86 032	12	28
33	9.83 821	13	9.97 801	25	0.02 199	9.86 020	12	27
34	9.83 834	14	9.97 826	25	0.02 174	9.86 008	12	26
35	9.83 848	13	9.97 851	26	0.02 149	9.85 996	12	25
36	9.83 861	13	9.97 877	25	0.02 123	9.85 984	12	24
37	9.83 874	13	9.97 902	25	0.02 098	9.85 972	12	23
38	9.83 887	14	9.97 927	26	0.02 073	9.85 960	12	22
39	9.83 901	13	9.97 953	25	0.02 047	9.85 948	12	21
40	9.83 914	13	9.97 978	25	0.02 022	9.85 936	12	20
41	9.83 927	13	9.98 003	26	0.01 997	9.85 924	12	19
42	9.83 940	14	9.98 029	25	0.01 971	9.85 912	12	18
43	9.83 954	13	9.98 054	25	0.01 946	9.85 900	12	17
44	9.83 967	13	9.98 079	25	0.01 921	9.85 888	12	16
45	9.83 980	13	9.98 104	26	0.01 896	9.85 876	12	15
46	9.83 993	13	9.98 130	25	0.01 870	9.85 864	13	14
47	9.84 006	14	9.98 155	25	0.01 845	9.85 851	12	13
48	9.84 020	13	9.98 180	26	0.01 820	9.85 839	12	12
49	9.84 033	13	9.98 206	25	0.01 794	9.85 827	12	11
50	9.84 046	13	9.98 231	25	0.01 769	9.85 815	12	10
51	9.84 059	13	9.98 256	25	0.01 744	9.85 803	12	9
52	9.84 072	13	9.98 281	26	0.01 719	9.85 791	12	8
53	9.84 085	13	9.98 307	25	0.01 693	9.85 779	13	7
54	9.84 098	14	9.98 332	25	0.01 668	9.85 766	12	6
55	9.84 112	13	9.98 357	26	0.01 643	9.85 754	12	5
56	9.84 125	13	9.98 383	25	0.01 617	9.85 742	12	4
57	9.84 138	13	9.98 408	25	0.01 592	9.85 730	12	3
58	9.84 151	13	9.98 433	25	0.01 567	9.85 718	12	2
59	9.84 164	13	9.98 458	26	0.01 542	9.85 706	13	1
60	9.84 177		9.98 484		0.01 516	9.85 693		0
	L. Cos.	d.	L. Cotg.	c.d.	L. Tang.	L. Sin.	d.	′

P. P.

	26	25
1	0,4	0,4
2	0,9	0,8
3	1,3	1,2
4	1,7	1,7
5	2,2	2,1
6	2,6	2,5
7	3,0	2,9
8	3,5	3,3
9	3,9	3,8
10	4,3	4,2
20	8,7	8,3
30	13,0	12,5
40	17,3	16,7
50	21,7	20,8

	14	13
1	0,2	0,2
2	0,5	0,4
3	0,7	0,6
4	0,9	0,9
5	1,2	1,1
6	1,4	1,3
7	1,6	1,5
8	1,9	1,7
9	2,1	2,0
10	2,3	2,2
20	4,7	4,3
30	7,0	6,5
40	9,3	8,7
50	11,7	10,8

	12	11
1	0,2	0,2
2	0,4	0,4
3	0,6	0,6
4	0,8	0,7
5	1,0	0,9
6	1,2	1,1
7	1,4	1,3
8	1,6	1,5
9	1,8	1,6
10	2,0	1,8
20	4,0	3,7
30	6,0	5,5
40	8,0	7,3
50	10,0	9,2

'	L. Sin.	d.	L. Tang.	c.d.	L. Cotg.	L. Cos.	d.	'	P. P.
0	9.84 177	18	9.98 484	25	0.01 516	9.85 698	12	60	
1	9.84 190	18	9.98 509	25	0.01 491	9.85 681	12	59	
2	9.84 208	18	9.98 534	26	0.01 466	9.85 669	12	58	
3	9.84 216	18	9.98 560	25	0.01 440	9.85 657	12	57	
4	9.84 229	18	9.98 585	25	0.01 415	9.85 645	13	56	
5	9.84 242	18	9.98 610	25	0.01 390	9.85 632	12	55	
6	9.84 255	14	9.98 635	26	0.01 365	9.85 620	12	54	
7	9.84 269	18	9.98 661	25	0.01 339	9.85 608	12	53	
8	9.84 282	18	9.98 686	25	0.01 314	9.85 596	13	52	
9	9.84 295	18	9.98 711	26	0.01 289	9.85 583	12	51	
10	9.84 308	18	9.98 737	25	0.01 263	9.85 571	12	50	26 25
11	9.84 321	18	9.98 762	25	0.01 238	9.85 559	12	49	
12	9.84 334	18	9.98 787	25	0.01 213	9.85 547	13	48	1 0,4 0,4
13	9.84 347	18	9.98 812	26	0.01 188	9.85 534	12	47	2 0,9 0,8
14	9.84 360	18	9.98 838	25	0.01 162	9.85 522	12	46	3 1,8 1,2
15	9.84 373	12	9.98 863	25	0.01 137	9.85 510	13	45	4 1,7 1,7
16	9.84 385	18	9.98 888	25	0.01 112	9.85 497	12	44	
17	9.84 398	18	9.98 913	26	0.01 087	9.85 485	12	43	5 2,2 2,1
18	9.84 411	18	9.98 939	25	0.01 061	9.85 473	13	42	6 2,6 2,5
19	9.84 424	18	9.98 964	25	0.01 036	9.85 460	12	41	7 3,0 2,9
									8 3,5 3,3
20	9.84 437	18	9.98 989	26	0.01 011	9.85 448	12	40	9 3,9 3,8
21	9.84 450	18	9.99 015	25	0.00 985	9.85 436	13	39	
22	9.84 463	18	9.99 040	25	0.00 960	9.85 423	12	38	10 4,3 4,2
23	9.84 476	18	9.99 065	25	0.00 935	9.85 411	12	37	20 8,7 8,3
24	9.84 489	18	9.99 090	26	0.00 910	9.85 399	13	36	30 13,0 12,5
25	9.84 502	18	9.99 116	25	0.00 884	9.85 386	12	35	40 17,3 16,7
26	9.84 515	18	9.99 141	25	0.00 859	9.85 374	13	34	50 21,7 20,8
27	9.84 528	12	9.99 166	25	0.00 834	9.85 361	12	33	
28	9.84 540	18	9.99 191	26	0.00 809	9.85 349	12	32	
29	9.84 553	18	9.99 217	25	0.00 783	9.85 337	13	31	
30	9.84 566	18	9.99 242	25	0.00 758	9.85 324	12	30	
31	9.84 579	18	9.99 267	26	0.00 733	9.85 312	13	29	
32	9.84 592	18	9.99 293	25	0.00 707	9.85 299	12	28	
33	9.84 605	18	9.99 318	25	0.00 682	9.85 287	13	27	
34	9.84 618	12	9.99 343	25	0.00 657	9.85 274	12	26	
35	9.84 630	18	9.99 368	26	0.00 632	9.85 262	12	25	14 13 12
36	9.84 643	18	9.99 394	25	0.00 606	9.85 250	13	24	
37	9.84 656	18	9.99 419	25	0.00 581	9.85 237	12	23	1 0,2 0,2 0,2
38	9.84 669	18	9.99 444	25	0.00 556	9.85 225	13	22	2 0,5 0,4 0,4
39	9.84 682	12	9.99 469	26	0.00 531	9.85 212	12	21	3 0,7 0,6 0,6
40	9.84 694	18	9.99 495	25	0.00 505	9.85 200	13	20	4 0,9 0,9 0,8
41	9.84 707	18	9.99 520	25	0.00 480	9.85 187	12	19	
42	9.84 720	18	9.99 545	25	0.00 455	9.85 175	13	18	5 1,2 1,1 1,0
43	9.84 733	12	9.99 570	26	0.00 430	9.85 162	12	17	6 1,4 1,3 1,2
44	9.84 745	18	9.99 596	25	0.00 404	9.85 150	13	16	7 1,6 1,5 1,4
									8 1,9 1,7 1,6
45	9.84 758	18	9.99 621	25	0.00 379	9.85 137	12	15	9 2,1 2,0 1,8
46	9.84 771	13	9.99 646	26	0.00 354	9.85 125	13	14	
47	9.84 784	12	9.99 672	25	0.00 328	9.85 112	13	13	10 2,3 2,2 2,0
48	9.84 796	18	9.99 697	25	0.00 303	9.85 100	12	12	20 4,7 4,3 4,0
49	9.84 809	13	9.99 722	25	0.00 278	9.85 087	13	11	30 7,0 6,5 6,0
50	9.84 822	13	9.99 747	26	0.00 253	9.85 074	12	10	40 9,3 8,7 8,0
51	9.84 835	12	9.99 773	25	0.00 227	9.85 062	13	9	50 11,7 10,8 10,0
52	9.84 847	13	9.99 798	25	0.00 202	9.85 049	12	8	
53	9.84 860	13	9.99 823	25	0.00 177	9.85 037	13	7	
54	9.84 873	12	9.99 848	26	0.00 152	9.85 024	12	6	
55	9.84 885	13	9.99 874	25	0.00 126	9.85 012	13	5	
56	9.84 898	13	9.99 899	25	0.00 101	9.84 999	13	4	
57	9.84 911	12	9.99 924	25	0.00 076	9.84 986	12	3	
58	9.84 923	13	9.99 949	26	0.00 051	9.84 974	13	2	
59	9.84 936	13	9.99 975	25	0.00 025	9.84 961	12	1	
60	9.84 949		0.00 000		0.00 000	9.84 949		0	
	L. Cos.	d.	L. Cotg.	c.d.	L. Tang.	L. Sin.	d.	'	P. P.

TABLE III.

THE NATURAL TRIGONOMETRIC FUNCTIONS.

For every ten minutes.

Sine.

°	0'	10'	20'	30'	40'	50'	60'		d.
0	0, 0000	0029	0058	0087	0116	0145	0175	89	29
1	0175	0204	0233	0262	0291	0320	0349	88	29
2	0349	0378	0407	0436	0465	0494	0523	87	29
3	0523	0552	0581	0610	0640	0669	0698	86	29
4	0698	0727	0756	0785	0814	0843	0872	85	29
5	0, 0872	0901	0929	0958	0987	1016	1045	84	29
6	1045	1074	1108	1132	1161	1190	1219	83	29
7	1219	1248	1276	1305	1334	1363	1392	82	29
8	1392	1421	1449	1478	1507	1536	1564	81	29
9	1564	1593	1622	1650	1679	1708	1736	80	29
10	0, 1736	1765	1794	1822	1851	1880	1908	79	29
11	1908	1937	1965	1994	2022	2051	2079	78	28
12	2079	2108	2136	2164	2193	2221	2250	77	28
13	2250	2278	2306	2334	2363	2391	2419	76	28
14	2419	2447	2476	2504	2532	2560	2588	75	28
15	0, 2588	2616	2644	2672	2700	2728	2756	74	28
16	2756	2784	2812	2840	2868	2896	2924	73	28
17	2924	2952	2979	3007	3035	3062	3090	72	28
18	3090	3118	3145	3173	3201	3228	3256	71	28
19	3256	3283	3311	3338	3365	3393	3420	70	27
20	0, 3420	3448	3475	3502	3529	3557	3584	69	27
21	3584	3611	3638	3665	3692	3719	3746	68	27
22	3746	3773	3800	3827	3854	3881	3907	67	27
23	3907	3934	3961	3987	4014	4041	4067	66	27
24	4067	4094	4120	4147	4173	4200	4226	65	26
25	0, 4226	4253	4279	4305	4331	4358	4384	64	26
26	4384	4410	4436	4462	4488	4514	4540	63	26
27	4540	4566	4592	4617	4643	4669	4695	62	26
28	4695	4720	4746	4772	4797	4823	4848	61	26
29	4848	4874	4899	4924	4950	4975	5000	60	25
30	0, 5000	5025	5050	5075	5100	5125	5150	59	25
31	5150	5175	5200	5225	5250	5275	5299	58	25
32	5299	5324	5348	5373	5398	5422	5446	57	24
33	5446	5471	5495	5519	5544	5568	5592	56	24
34	5592	5616	5640	5664	5688	5712	5736	55	24
35	0, 5736	5760	5783	5807	5831	5854	5878	54	24
36	5878	5901	5925	5948	5972	5995	6018	53	23
37	6018	6041	6065	6088	6111	6134	6157	52	23
38	6157	6180	6202	6225	6248	6271	6293	51	23
39	6293	6316	6338	6361	6383	6406	6428	50	22
40	0, 6428	6450	6472	6494	6517	6539	6561	49	22
41	6561	6583	6604	6626	6648	6670	6691	48	22
42	6691	6713	6734	6756	6777	6799	6820	47	22
43	6820	6841	6862	6884	6905	6926	6947	46	21
44	6947	6967	6988	7009	7030	7050	7071	45	21
45	0, 7071								
	60'	50'	40'	30'	20'	10'	0'	°	d.

P. P.

	30	29
1	3,0	2,9
2	6,0	5,8
3	9,0	8,7
4	12,0	11,6
5	15,0	14,5
6	18,0	17,4
7	21,0	20,3
8	24,0	23,2
9	27,0	26,1

	28	27
1	2,8	2,7
2	5,6	5,4
3	8,4	8,1
4	11,2	10,8
5	14,0	13,5
6	16,8	16,2
7	19,6	18,9
8	22,4	21,6
9	25,2	24,3

	26	25
1	2,6	2,5
2	5,2	5,0
3	7,8	7,5
4	10,4	10,0
5	13,0	12,5
6	15,6	15,0
7	18,2	17,5
8	20,8	20,0
9	23,4	22,5

	24	23
1	2,4	2,3
2	4,8	4,6
3	7,2	6,9
4	9,6	9,2
5	12,0	11,5
6	14,4	13,8
7	16,8	16,1
8	19,2	18,4
9	21,6	20,7

	22	21	20
1	2,2	2,1	2,0
2	4,4	4,2	4,0
3	6,6	6,3	6,0
4	8,8	8,4	8,0
5	11,0	10,5	10,0
6	13,2	12,6	12,0
7	15,4	14,7	14,0
8	17,6	16,8	16,0
9	19,8	18,9	18,0

Cosine.

°	0'	10'	20'	80'	40'	50'	60'		d.
45	0,7071	7092	7112	7188	7158	7173	7193	44	20
46	7193	7214	7284	7254	7274	7294	7314	43	20
47	7314	7383	7358	7373	7392	7412	7431	42	20
48	7431	7451	7470	7490	7509	7528	7547	41	19
49	7547	7566	7585	7604	7623	7642	7660	40	19
50	0,7660	7679	7698	7716	7735	7753	7771	39	18
51	7771	7790	7808	7826	7844	7862	7880	38	18
52	7880	7898	7916	7934	7951	7969	7986	37	18
53	7986	8004	8021	8039	8056	8073	8090	36	17
54	8090	8107	8124	8141	8158	8175	8192	35	17
55	0,8192	8208	8225	8241	8258	8274	8290	34	16
56	8290	8307	8323	8339	8355	8371	8387	33	16
57	8387	8403	8418	8434	8450	8465	8480	32	16
58	8480	8496	8511	8526	8542	8557	8572	31	15
59	8572	8587	8601	8616	8631	8646	8660	30	15
60	0,8660	8675	8689	8704	8718	8732	8746	29	14
61	8746	8760	8774	8788	8802	8816	8829	28	14
62	8829	8843	8857	8870	8884	8897	8910	27	14
63	8910	8923	8936	8949	8962	8975	8988	26	13
64	8988	9001	9013	9026	9038	9051	9063	25	12
65	0,9063	9075	9088	9100	9112	9124	9135	24	12
66	9135	9147	9159	9171	9182	9194	9205	23	12
67	9205	9216	9228	9239	9250	9261	9272	22	11
68	9272	9288	9298	9304	9315	9325	9336	21	11
69	9336	9346	9356	9367	9377	9387	9397	20	10
70	0,9397	9407	9417	9426	9436	9446	9455	19	10
71	9455	9465	9474	9483	9492	9502	9511	18	9
72	9511	9520	9528	9537	9546	9555	9563	17	9
73	9563	9572	9580	9588	9596	9605	9613	16	8
74	9613	9621	9628	9636	9644	9652	9659	15	8
75	0,9659	9667	9674	9681	9689	9696	9703	14	7
76	9703	9710	9717	9724	9730	9737	9744	13	7
77	9744	9750	9757	9763	9769	9775	9781	12	6
78	9781	9787	9793	9799	9805	9811	9816	11	6
79	9816	9822	9827	9833	9838	9843	9848	10	5
80	0,9848	9853	9858	9863	9868	9872	9877	9	5
81	9877	9881	9886	9890	9894	9899	9903	8	4
82	9903	9907	9911	9914	9918	9922	9925	7	4
83	9925	9929	9932	9936	9939	9942	9945	6	3
84	9945	9948	9951	9954	9957	9959	9962	5	3
85	0,9962	9964	9967	9969	9971	9974	9976	4	2
86	9976	9978	9980	9981	9983	9985	9986	3	2
87	9986	9988	9989	9990	9992	9993	9994	2	1
88	9994	9995	9996	9997	9997	9998	9998	1	1
89	9998	9999	9999	*0000	*0000	*0000	*0000	0	0
90	1,0000								
	60'	50'	40'	80'	20'	10'	0'	°	d.

P. P.

	21	20	19	18
1	2,1	2,0	1,9	1,8
2	4,2	4,0	3,8	3,6
3	6,3	6,0	5,7	5,4
4	8,4	8,0	7,6	7,2
5	10,5	10,0	9,5	9,0
6	12,6	12,0	11,4	10,8
7	14,7	14,0	13,3	12,6
8	16,8	16,0	15,2	14,4
9	18,9	18,0	17,1	16,2

	17	16	15	14
1	1,7	1,6	1,5	1,4
2	3,4	3,2	3,0	2,8
3	5,1	4,8	4,5	4,2
4	6,8	6,4	6,0	5,6
5	8,5	8,0	7,5	7,0
6	10,2	9,6	9,0	8,4
7	11,9	11,2	10,5	9,8
8	13,6	12,8	12,0	11,2
9	15,3	14,4	13,5	12,6

	13	12	11	10
1	1,3	1,2	1,1	1,0
2	2,6	2,4	2,2	2,0
3	3,9	3,6	3,3	3,0
4	5,2	4,8	4,4	4,0
5	6,5	6,0	5,5	5,0
6	7,8	7,2	6,6	6,0
7	9,1	8,4	7,7	7,0
8	10,4	9,6	8,8	8,0
9	11,7	10,8	9,9	9,0

	9	8	7	6
1	0,9	0,8	0,7	0,6
2	1,8	1,6	1,4	1,2
3	2,7	2,4	2,1	1,8
4	3,6	3,2	2,8	2,4
5	4,5	4,0	3,5	3,0
6	5,4	4,8	4,2	3,6
7	6,3	5,6	4,9	4,2
8	7,2	6,4	5,6	4,8
9	8,1	7,2	6,3	5,4

	5	4	3	2
1	0,5	0,4	0,3	0,2
2	1,0	0,8	0,6	0,4
3	1,5	1,2	0,9	0,6
4	2,0	1,6	1,2	0,8
5	2,5	2,0	1,5	1,0
6	3,0	2,4	1,8	1,2
7	3,5	2,8	2,1	1,4
8	4,0	3,2	2,4	1,6
9	4,5	3,6	2,7	1,8

P. P.

Cosine.

Tangent.

°	0'	10'	20'	30'	40'	50'	60'		d.
0	0, 0000	0029	0058	0087	0116	0145	0175	89	29
1	0175	0204	0233	0262	0291	0320	0349	88	29
2	0349	0378	0407	0437	0466	0495	0524	87	29
3	0524	0553	0582	0612	0641	0670	0699	86	29
4	0699	0729	0758	0787	0816	0846	0875	85	29
5	0, 0875	0904	0934	0963	0992	1022	1051	84	29
6	1051	1080	1110	1139	1169	1198	1228	83	30
7	1228	1257	1287	1317	1346	1376	1405	82	30
8	1405	1435	1465	1495	1524	1554	1584	81	30
9	1584	1614	1644	1673	1703	1733	1763	80	30
10	0, 1763	1798	1823	1853	1883	1914	1944	79	30
11	1944	1974	2004	2035	2065	2095	2126	78	30
12	2126	2156	2186	2217	2247	2278	2309	77	30
13	2309	2339	2370	2401	2432	2462	2493	76	31
14	2493	2524	2555	2586	2617	2648	2679	75	31
15	0, 2679	2711	2742	2773	2805	2836	2867	74	31
16	2867	2899	2931	2962	2994	3026	3057	73	32
17	3057	3089	3121	3153	3185	3217	3249	72	32
18	3249	3281	3314	3346	3378	3411	3443	71	32
19	3443	3476	3508	3541	3574	3607	3640	70	33
20	0, 3640	3673	3706	3739	3772	3805	3839	69	33
21	3839	3872	3906	3939	3973	4006	4040	68	34
22	4040	4074	4108	4142	4176	4210	4245	67	34
23	4245	4279	4314	4348	4383	4417	4452	66	34
24	4452	4487	4522	4557	4592	4628	4663	65	35
25	0, 4663	4699	4734	4770	4806	4841	4877	64	36
26	4877	4913	4950	4986	5022	5059	5095	63	36
27	5095	5132	5169	5206	5243	5280	5317	62	37
28	5317	5354	5392	5430	5467	5505	5543	61	38
29	5543	5581	5619	5658	5696	5735	5774	60	38
30	0, 5774	5812	5851	5890	5930	5969	6009	59	39
31	6009	6048	6088	6128	6168	6208	6249	58	40
32	6249	6289	6330	6371	6412	6453	6494	57	41
33	6494	6536	6577	6619	6661	6703	6745	56	42
34	6745	6787	6830	6873	6916	6959	7002	55	43
35	0, 7002	7046	7089	7133	7177	7221	7265	54	44
36	7265	7310	7355	7400	7445	7490	7536	53	45
37	7536	7581	7627	7673	7720	7766	7813	52	46
38	7813	7860	7907	7954	8002	8050	8098	51	48
39	8098	8146	8195	8243	8292	8342	8391	50	49
40	0, 8391	8441	8491	8541	8591	8642	8693	49	50
41	8693	8744	8796	8847	8899	8952	9004	48	52
42	9004	9057	9110	9163	9217	9271	9325	47	54
43	9325	9380	9435	9490	9545	9601	9657	46	55
44	9657	9718	9770	9827	9884	9942	₌0000	45	57
45	1, 0000								
	60'	50'	40'	30'	20'	10'	0'	°	d.

P. P.

	29	30	31	32	33
1	2,9	3,0	3,1	3,2	3,3
2	5,8	6,0	6,2	6,4	6,6
3	8,7	9,0	9,3	9,6	9,9
4	11,6	12,0	12,4	12,8	13,2
5	14,5	15,0	15,5	16,0	16,5
6	17,4	18,0	18,6	19,2	19,8
7	20,3	21,0	21,7	22,4	23,1
8	23,2	24,0	24,8	25,6	26,4
9	26,1	27,0	27,9	28,8	29,7

	34	35	36	37	38
1	3,4	3,5	3,6	3,7	3,8
2	6,8	7,0	7,2	7,4	7,6
3	10,2	10,5	10,8	11,1	11,4
4	13,6	14,0	14,4	14,8	15,2
5	17,0	17,5	18,0	18,5	19,0
6	20,4	21,0	21,6	22,2	22,8
7	23,8	24,5	25,2	25,9	26,6
8	27,2	28,0	28,8	29,6	30,4
9	30,6	31,5	32,4	33,3	34,2

	39	40	41	42	43
1	3,9	4,0	4,1	4,2	4,3
2	7,8	8,0	8,2	8,4	8,6
3	11,7	12,0	12,3	12,6	12,9
4	15,6	16,0	16,4	16,8	17,2
5	19,5	20,0	20,5	21,0	21,5
6	23,4	24,0	24,6	25,2	25,8
7	27,3	28,0	28,7	29,4	30,1
8	31,2	32,0	32,8	33,6	34,4
9	35,1	36,0	36,9	37,8	38,7

	44	45	46	47	48
1	4,4	4,5	4,6	4,7	4,8
2	8,8	9,0	9,2	9,4	9,6
3	13,2	13,5	13,8	14,1	14,4
4	17,6	18,0	18,4	18,8	19,2
5	22,0	22,5	23,0	23,5	24,0
6	26,4	27,0	27,6	28,2	28,8
7	30,8	31,5	32,2	32,9	33,6
8	35,2	36,0	36,8	37,6	38,4
9	39,6	40,5	41,4	42,3	43,2

	49	50	51	52	53
1	4,9	5,0	5,1	5,2	5,3
2	9,8	10,0	10,2	10,4	10,6
3	14,7	15,0	15,3	15,6	15,9
4	19,6	20,0	20,4	20,8	21,2
5	24,5	25,0	25,5	26,0	26,5
6	29,4	30,0	30,6	31,2	31,8
7	34,3	35,0	35,7	36,4	37,1
8	39,2	40,0	40,8	41,6	42,4
9	44,1	45,0	45,9	46,8	47,7

	54	55	56	57	58
1	5,4	5,5	5,6	5,7	5,8
2	10,8	11,0	11,2	11,4	11,6
3	16,2	16,5	16,8	17,1	17,4
4	21,6	22,0	22,4	22,8	23,2
5	27,0	27,5	28,0	28,5	29,0
6	32,4	33,0	33,6	34,2	34,8
7	37,8	38,5	39,2	39,9	40,6
8	43,2	44,0	44,8	45,6	46,4
9	48,6	49,5	50,4	51,3	52,2

P. P.

Cotangent.

Tangent.

°	0'	10'	20'	30'	40'	50'	60'	d.
45	1,000	1,006	1,012	1,018	1,024	1,030	1,086	44
46	1,086	1,042	1,048	1,054	1,060	1,066	1,072	43
47	1,072	1,079	1,085	1,091	1,098	1,104	1,111	42
48	1,111	1,117	1,124	1,130	1,137	1,144	1,150	41
49	1,150	1,157	1,164	1,171	1,178	1,185	1,192	40
50	1,192	1,199	1,206	1,213	1,220	1,228	1,235	39
51	1,235	1,242	1,250	1,257	1,265	1,272	1,280	38
52	1,280	1,288	1,295	1,303	1,311	1,319	1,327	37
53	1,327	1,335	1,343	1,351	1,360	1,368	1,376	36
54	1,376	1,385	1,393	1,402	1,411	1,419	1,428	35
55	1,428	1,437	1,446	1,455	1,464	1,473	1,483	34
56	1,483	1,492	1,501	1,511	1,520	1,530	1,540	33
57	1,540	1,550	1,560	1,570	1,580	1,590	1,600	32
58	1,600	1,611	1,621	1,632	1,643	1,653	1,664	31
59	1,664	1,675	1,686	1,698	1,709	1,720	1,732	30
60	1,732	1,744	1,756	1,767	1,780	1,792	1,804	29
61	1,804	1,816	1,829	1,842	1,855	1,868	1,881	28
62	1,881	1,894	1,907	1,921	1,935	1,949	1,963	27
63	1,963	1,977	1,991	2,006	2,020	2,035	2,050	26
64	2,050	2,066	2,081	2,097	2,112	2,128	2,145	25
65	2,145	2,161	2,177	2,194	2,211	2,229	2,246	24
66	2,246	2,264	2,282	2,300	2,318	2,337	2,356	23
67	2,356	2,375	2,394	2,414	2,434	2,455	2,475	22
68	2,475	2,496	2,517	2,539	2,560	2,583	2,605	21
69	2,605	2,628	2,651	2,675	2,699	2,723	2,747	20
70	2,747	2,773	2,798	2,824	2,850	2,877	2,904	19
71	2,904	2,932	2,960	2,989	3,018	3,047	3,078	18
72	3,078	3,108	3,140	3,172	3,204	3,237	3,271	17
73	3,271	3,305	3,340	3,376	3,412	3,450	3,487	16
74	3,487	3,526	3,566	3,606	3,647	3,689	3,732	15
75	3,732	3,776	3,821	3,867	3,914	3,962	4,011	14
76	4,011	4,061	4,113	4,165	4,219	4,275	4,331	13
77	4,331	4,390	4,449	4,511	4,574	4,638	4,705	12
78	4,705	4,773	4,843	4,915	4,989	5,066	5,145	11
79	5,145	5,226	5,309	5,396	5,485	5,576	5,671	10
80	5,671	5,769	5,871	5,976	6,084	6,197	6,314	9
81	6,314	6,435	6,561	6,691	6,827	6,968	7,115	8
82	7,115	7,269	7,429	7,596	7,770	7,953	8,144	7
83	8,144	8,345	8,556	8,777	9,010	9,255	9,514	6
84	9,514	9,788	10,078	10,385	10,712	11,059	11,430	5
85	11,480	11,826	12,251	12,706	13,197	13,727	14,301	4
86	14,301	14,924	15,605	16,350	17,169	18,075	19,081	3
87	19,081	20,206	21,470	22,904	24,542	26,432	28,636	2
88	28,636	31,242	34,368	38,188	42,964	49,104	57,290	1
89	57,290	68,750	85,940	114,59	171,89	343,77	infinit.	0
90	infinit.							

P. P.

	6	7	8	9	10
1	0,6	0,7	0,8	0,9	1,0
2	1,2	1,4	1,6	1,8	2,0
3	1,8	2,1	2,4	2,7	3,0
4	2,4	2,8	3,2	3,6	4,0
5	3,0	3,5	4,0	4,5	5,0
6	3,6	4,2	4,8	5,4	6,0
7	4,2	4,9	5,6	6,3	7,0
8	4,8	5,6	6,4	7,2	8,0
9	5,4	6,3	7,2	8,1	9,0

	12	13	14	15
1	1,2	1,3	1,4	1,5
2	2,4	2,6	2,8	3,0
3	3,6	3,9	4,2	4,5
4	4,8	5,2	5,6	6,0
5	6,0	6,5	7,0	7,5
6	7,2	7,8	8,4	9,0
7	8,4	9,1	9,8	10,5
8	9,6	10,4	11,2	12,0
9	10,8	11,7	12,6	13,5

	17	18	19	20
1	1,7	1,8	1,9	2,0
2	3,4	3,6	3,8	4,0
3	5,1	5,4	5,7	6,0
4	6,8	7,2	7,6	8,0
5	8,5	9,0	9,5	10,0
6	10,2	10,8	11,4	12,0
7	11,9	12,6	13,3	14,0
8	13,6	14,4	15,2	16,0
9	15,3	16,2	17,1	18,0

	22	23	24	25
1	2,2	2,3	2,4	2,5
2	4,4	4,6	4,8	5,0
3	6,6	6,9	7,2	7,5
4	8,8	9,2	9,6	10,0
5	11,0	11,5	12,0	12,5
6	13,2	13,8	14,4	15,0
7	15,4	16,1	16,8	17,5
8	17,6	18,4	19,2	20,0
9	19,8	20,7	21,6	22,5

	27	28	59	62
1	2,7	2,8	5,9	6,2
2	5,4	5,6	11,8	12,4
3	8,1	8,4	17,7	18,6
4	10,8	11,2	23,6	24,8
5	13,5	14,0	29,5	31,0
6	16,2	16,8	35,4	37,2
7	18,9	19,6	41,3	43,4
8	21,6	22,4	47,2	49,6
9	24,3	25,2	53,1	55,8

	64	67	68	70
1	6,4	6,7	6,8	7,0
2	12,8	13,4	13,6	14,0
3	19,2	20,1	20,4	21,0
4	25,6	26,8	27,2	28,0
5	32,0	33,5	34,0	35,0
6	38,4	40,2	40,8	42,0
7	44,8	46,9	47,6	49,0
8	51,2	53,6	54,4	56,0
9	57,6	60,3	61,2	63,0

60'	50'	40'	30'	20'	10'	0'	°	d.	P. P.

Cotangent.

ADVERTISEMENTS

TREATISE ON TRIGONOMETRY

AND ITS APPLICATIONS TO

ASTRONOMY AND GEODESY

BY EDWARD A. BOWSER, LL.D.

Professor of Mathematics and Engineering in Rutgers College

The aim of the author has been to present in as concise a form as is consistent with clearness, the fullest course in Trigonometry which is given in the best technical schools and in advanced courses in colleges.

The examples are very numerous and are carefully selected. Among these are some of the most elegant theorems in Plane and Spherical Trigonometry. The numerical solution of triangles has received much attention, each case being treated in detail.

The chapters on De Moivre's Theorem, and Astronomy, Geodesy, and Polyhedrons will serve to introduce the students to some of the higher applications of Trigonometry, rarely found in American text-books.

American Mathematical Monthly : Excepting one, this is the most complete Treatise on Trigonometry published in America, and in point of excellence is superior to that work. In the method of treatment, arrangement, typographical execution, and numerous and well-selected exercises, it has no superior. The definitions of the functions are given "once for all" and need not be restated and modified when obtuse and reflex angles are considered.

In the development of the theoretical part of the subject, the work is especially interesting and clear. From the beginning the student is carried along with enthusiasm and with the assurance that he is mastering the subject. The unusually large and well-chosen collection of problems are suited to every requirement, and by solving these the student learns to do by doing.

The treatment of Trigonometric Elimination, De Moivre's Theorem, Summation of Series, etc., is more complete than is usually given in text-books.

These observations have been gathered by using the book in the class-room.

Half leather. Pages, xiv + 368. Introduction price, $1.50.

Bowser's Five Place Logarithmic and Trigonometric Tables, 50 cents.
Bowser's Elements of Plane and Spherical Trigonometry, 90 cents.
   With tables, $1.40.
Bowser's Plane and Solid Geometry, $1.25.
Bowser's Academic Algebra, $1.12.
Bowser's College Algebra, $1.50.

D. C. HEATH & CO., Publishers, Boston, New York, Chicago

THEORY OF EQUATIONS

By SAMUEL MARX BARTON, Ph.D.,

Professor of Mathematics in the University of the South.

In this treatise the author aims to give the elements of Determinants and the Theory of Equations in a form suitable, both in amount and quality of matter, for use in undergraduate courses. The work is readily intelligible to the average student who has become proficient in algebra and the elements of trigonometry. The use of the calculus has been purposely avoided. While the presentation of the subject has necessarily been condensed to suit the requirements of college courses, great pains has been taken not to sacrifice clearness to brevity. It is a short treatise. but not a syllabus.

Part I treats of Determinants. The chapters give the fundamental theorems, with examples for illustration; applications and special forms of determinants, followed by a collection of carefully selected examples.

Part II treats of the Theory of Equations proper, with chapters upon complex numbers, properties of polynomials, general properties of equations, relations between roots and coefficients, symmetric functions, transformation of equations, limits of the roots of an equation, separation of roots, elimination, solution of numerical equations. Almost every theorem is elucidated by the complete solution of one or more representative examples.

Cloth. Pages, x + 198. Introduction price, $1.50.

D. C. HEATH & CO., Publishers, Boston, New York, Chicago

COLLEGE ALGEBRA

By EDWARD A. BOWSER, LL.D.

Professor of Mathematics and Engineering in Rutgers College.

This work is designed for academies, colleges and scientific schools. It begins with the elements, and the full treatment of the earlier parts renders it unnecessary that students who use it shall have previously studied a more elementary algebra.

Among its points of superiority are the following : —

1. Completeness of treatment combined with simplicity.

2. Avoidance of the abstruse and the elaborate in treating the more difficult parts of the subject.

3. Definiteness of statement — the steps and processes are generally formulated in plain rules.

4. Careful consideration and clear presentation of material for the student.

5. Systematic arrangement of material under each subject.

6. Full notes of explanation, direction, and information, useful to student and teacher.

7. Numerous examples are distributed throughout the text in immediate connection with the principles they illustrate.

Half leather. Pages, xviii + 540. Introduction price, $1.50.

Bowser's Academic Algebra, $1.12.
Bowser's Plane and Solid Geometry, $1.25.
Bowser's Elements of Place and Spherical Trigonometry, 90 cents.
    With tables, $1.40.
Bowser's Five Place Logarithmic and Trigonometric Tables, 50 cents.
Bowser's Treatise on Trigonometry, $1.50.

D. C. HEATH & CO., Publishers, Boston, New York, Chicago

ANALYTIC GEOMETRY

PLANE AND SOLID.

BY E. W. NICHOLS,

Professor of Mathematics in the Virginia Military Institute.

The aim of the author has been to prepare a work for beginners, and at the same time to make it sufficiently comprehensive for the requirements of the usual undergraduate course. For the methods of development of the various principles he has drawn largely upon his experience in the classroom. In the preparation of the work, all authors, home and foreign, whose works were available, have been freely consulted.

In the first few chapters elementary examples follow the discussion of each principle. In the subsequent chapters, sets of examples appear at intervals throughout each chapter, and are so arranged as to partake both of the nature of a review and an extension of the preceding principles. At the end of each chapter general examples, involving a more extended application of the principles deduced, are placed for the benefit of those who may desire a higher course in the subject.

Nichols's Analytic Geometry is in use as the regular text in the greater number of the larger colleges and universities, and has proved itself adapted to the needs of institutions with the most varied requirements.

Cloth. Pages xii + 275. Introduction price, $1.25.

D. C. HEATH & CO., Publishers, Boston, New York, Chicago

DIFFERENTIAL
AND INTEGRAL CALCULUS

BY E. W. NICHOLS,

*Professor of Mathematics in the Virginia Military Institute,
and author of "Nichols' Analytic Geometry."*

The author, after twenty years' experience in the class-room as professor of applied and pure mathematics, has embodied the results of his experience in this work.

The principal features of the text are these:

1. The work is limited in scope to the usual undergraduate course given in colleges, universities and technical schools.

2. The treatment is based on the "Methods of Rates and Limits," or the "Method of Limits" alone. Where a brief course is desired, the latter method may be pursued without destroying the continuity of the subject.

3. Historical notes at the heads of chapters give a brief account of the discovery and development of the subjects treated. Foot-notes refer to problems of particular historic interest.

4. The principles are extensively applied to geometric, mechanical, electrical and engineering problems, to arouse the interest of the student and impress principles by use. These problems prepare for an intelligent study of applied mathematics.

5. A chapter on differential equations is added for those who may desire an elementary knowledge of this interesting extension of the subject. The chapter is limited to the ordinary principles required by students in mathematical physics.

Cloth. Pages, 408. Price, $2.00.

D. C. HEATH & CO., Publishers, Boston, New York, Chicago

EXERCISE BOOK IN ALGEBRA

Designed for supplementary or review work in connection with any text-book in Algebra.

By MATTHEW S. McCURDY, M.A.,

Instructor in Mathematics in the Phillips Academy, Andover, Mass.

This book is designed to furnish a collection of exercises similar in character to those in the ordinary text-books, of medium grade as to difficulty, and selected with special reference to giving an opportunity for drill upon those subjects which experience has shown to be difficult for students to master.

Though intended primarily to be supplementary to some regular text-book, a number of definitions and a few rules have been added, in the hope that it may also be found useful as an independent review and drill book. With or without answers.

Cloth. Pages, vi + 220. Introduction price, 60 cents.

ALGEBRA LESSONS
By J. H. GILBERT.

This series is intended for supplementary or review work, and contains three numbers: No. 1 — To Fractional Equations, No. 2 — Through Quadratic Equations, No. 3 — Higher Algebra.

Paper. Tablet form. Price, $1.44 per dozen.

REVIEW AND TEST PROBLEMS IN ALGEBRA
By S. J. PETERSON AND L. F. BALDWIN.

The problems in this manual are original — none have been copied from any other author. They illustrate points of special importance, and are sufficiently varied and difficult for written drills for those preparing for college entrance examinations.

Paper. 87 pages. Introduction price, 30 cents.

D. C. HEATH & CO., Publishers, Boston, New York, Chicago

Mathematics

Barton's Plane Surveying. With tables and a chapter on levelling. $1.30.

Barton's Theory of Equations. A treatise for college classes. $1.50.

Bauer and Brooke's Plane and Spherical Trigonometry. $1.50.

Bowser's Academic Algebra. For secondary schools. $1.12.

Bowser's College Algebra. A full treatment of elementary and advanced topics. $1.50

Bowser's Plane and Solid Geometry. $1.25. Plane, bound separately. 75 cts.

Bowser's Elements of Plane and Spherical Trigonometry. 90 cts.; with tables, $1.40.

Bowser's Treatise on Plane and Spherical Trigonometry. $1.50.

Bowser's Five-Place Logarithmic Tables. 50 cts.

Candy's Analytic Geometry. Plane and Solid. $1.50.

Cohen's Differential Equations. An elementary treatise. $2.00.

Fine's Number System of Algebra. Theoretical and historical. $1.00.

Gilbert's Algebra Lessons. Three numbers: No. 1, to Fractional Equations; No. 2, through Quadratic Equations; No. 3, Higher Algebra. Each number, per dozen, $1.44.

Hopkins's Plane Geometry. Follows the inductive method. 75 cts.

Howland's Elements of the Conic Sections. 75 cts.

Lefevre's Number and its Algebra. Introductory to college courses in Algebra. $1.25.

Lyman's Geometry Exercises. Supplementary work for drill. Per dozen, $1.60.

McCurdy's Exercise Book in Algebra. A thorough drill book. 60 cts.

Nichols's Analytic Geometry. A treatise for college courses. $1.25.

Nichols's Calculus. Differential and Integral. $2.00.

Osborne's Differential and Integral Calculus. Revised edition. $2.00.

Peterson and Baldwin's Problems in Algebra. For texts and reviews. 30 cts.

Robbins's Surveying and Navigation. A brief and practical treatise. 50 cts.

Waldo's Descriptive Geometry. With problems and suggestions. 80 cts.

Wells's Academic Arithmetic. With or without answers. $1.00.

Wells's Algebra for Secondary Schools. $1.20. Pocket edition, $1.20.

Wells's Essentials of Algebra. For secondary schools. $1.10.

Wells's Academic Algebra. With or without answers. $1.08.

Wells's Text Book in Algebra. $1.40.

Wells's New Higher Algebra. For schools and colleges. $1.32.

Wells's Higher Algebra. $1.32.

Wells's University Algebra. Octavo. $1.50.

Wells's College Algebra. $1.50. Part II, beginning with quadratics. $1.32.

Wells's Advanced Course in Algebra. A college algebra. $1.50.

Wells's Essentials of Geometry. $1.25. Plane, 75 cts. Solid, 75 cts.

Wells's New Plane and Spherical Trigonometry. For colleges and technical schools. $1.00. With six-place tables, $1.25. With Robbins's Surveying and Navigation, $1.50.

Wells's Complete Trigonometry. Plane and Spherical. 90 cts. With tables, $1.08. Plane, bound separately, 75 cts.

Wells's New Six-Place Logarithmic Tables. 60 cts. Pocket edition, small type, 36 cts.

Wells's Four-Place Tables. 25 cts.

Wright's Exercises in Concrete Geometry. 30 cts.

For Arithmetics see our list of books in Elementary Mathematics.

D. C. HEATH & CO., Publishers, Boston, New York, Chicago

Heath's Pedagogical Library.

Sent postpaid on receipt of price by the publishers.
Special catalogue, with full descriptions, free on request.

D. C. HEATH & CO., Publishers, Boston, New York, Chicago

i
ry

he

rk

th

m

d

age 46.
odd. numbers
3 prob.

No
Dec
≤

35

$C = 18$

$b = a$

DUE OCT 27 '44

$C = \underline{a}$

given by;
required

$\overline{a} = 36.7$

$\overline{C} = 360\,5$

$\dfrac{t}{\overline{\sin A}}$

$B \cdot 72^\circ 5$

$e \quad 71^\circ 5$

$A\ 58.184$

$C\ 57.801$

$\overline{}\log 1.7863$

$\log \sin .07559$